Barbara Merchant

PLUMBING

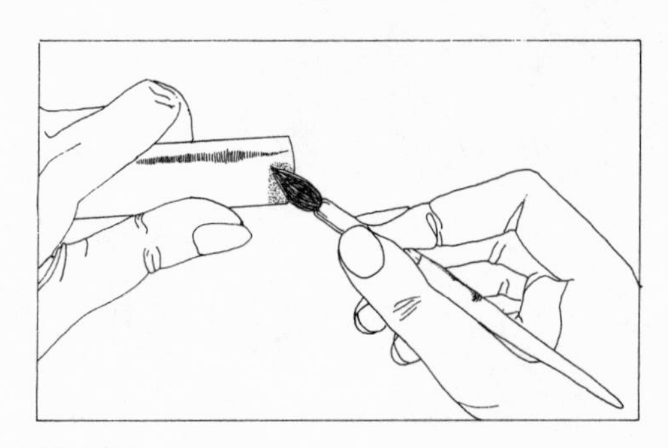

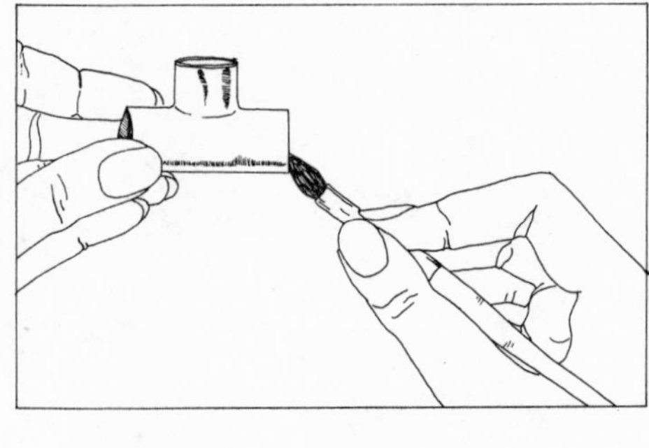

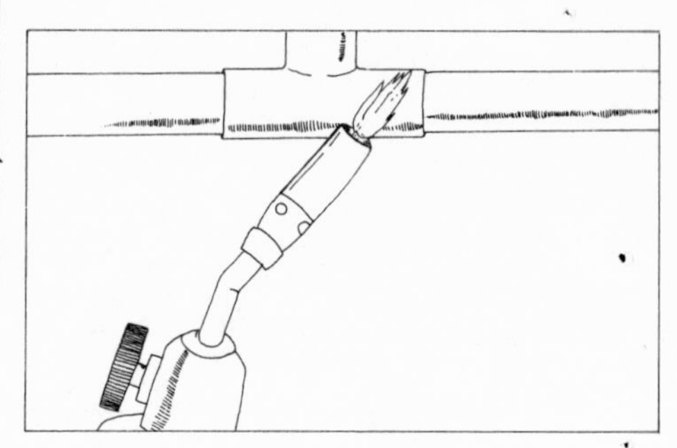

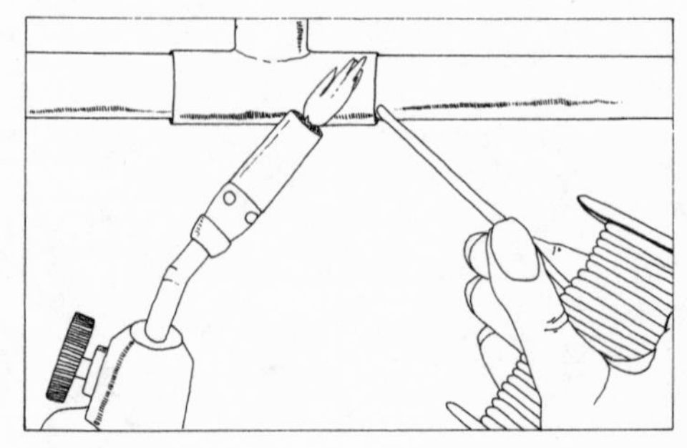

FOR DUMMIES

by Don Fredriksson

Illustrations by HOLLY EVANS SAMMONS

THE BOBBS-MERRILL COMPANY, INC.

Indianapolis/New York

Published by The Bobbs-Merrill Co., Inc.
Indianapolis/New York
Manufactured in the United States of America
First Printing
Designed by Chris Simon

Library of Congress Cataloging in Publication Data
Fredriksson, Don.
Plumbing for dummies.
Includes index.
1. Plumbing—Amateurs' manuals. I. Title.
TH6124.F74 1983 643'.6'0288 82-17790
ISBN 0-672-52738-3

For Sherry, my wife; Kathy Poropat, my agent; the people at the National Plumbing Information Service, and Tom Higham and his International Association of Plumbing and Mechanical Officials, all of whom made this volume possible

Contents

List of Illustrations

Many old-time journeyman plumbers used to define a dummy as a first-year apprentice; e.g. "Hey, dummy! Gimme a 18" Stillson," or "Tell the dummy to get a pail of beer from Dougan's." But "dummies" eventually became journeymen and masters. Perhaps this book will make your road easier than mine has been, whether you're a bus driver, secretary, the cop on the beat, or my travel agent down the street.

My hat's off to you. Good luck!

Introduction

For over twenty years, I have been remedying the plumbing errors and disasters of friends, acquaintances, and the clients of my plumbing shops in both the eastern and western United States.

Recently, there have been more serious problems of these types to fix, including some which have been created as a result of people reading do-it-yourself plumbing books that do not faithfully follow the *Uniform Plumbing Code* (1979 Edition as amended) which is valid throughout much of the English-speaking world.

This authoritative code has been officially recognized in the following American states and jurisdictions:

Alaska*
Arizona
California*
Colorado
Delaware
Hawaii*
Idaho*
Illinois
Indiana*
Iowa*
Kansas
Maine*
Michigan
Minnesota
Missouri
Montana*
Nebraska
Nevada*
New Mexico*
North Dakota
Ohio
Oklahoma
Oregon*
Pennsylvania
South Dakota
Texas
Utah*
Vermont
Washington*
West Virginia
Wyoming
Island of Guam*

The asterisk (*) indicates that the *UPC* has been adopted as the state code. Those states not so marked have towns, townships, cities, or counties using the *UPC*.

New York, New Jersey, Massachusetts, and the District of Columbia all have their own plumbing codes. The southeastern states recognize the *Southern Standard Plumbing Code*.

Other jurisdictions within the United States use the *UPC*, but have not listed with the International Association of Plumbing and Mechanical Officials (IAPMO).

Outside of the United States, the *UPC* is recognized in: Germany, Japan, Saudi Arabia, Guatemala, and Trinidad. The *UPC* may not be used exclusively in those countries, but IAPMO has been told that some jurisdictions within their borders are using it either as the code or as a guide.

It is sufficient to realize that, from a population standpoint, the *UPC* is the basis for more plumbing inspections than is any other code in the country. If you follow its provisions as a basic guide, you'll be in the right ballpark nearly everywhere in the United States.

However, if your locality does not recognize the *UPC*, I can't stress too strongly the importance of getting a copy of your local code and checking it against the information and data in this book. Chances are, the combination of this guide and your local code will do the job for you nicely. Probably, the only changes will be pipe and distance specifications.

Recently, I had the occasion to compare a typical do-it-yourself plumbing book full of pretty color pictures to the *UPC* in California. This so-called manual said that a water closet (toilet) may be installed seven feet from the vent which serves it. The 1979 edition of the *UPC* plainly states (page 56, section 702) that the maximum distance cannot be more than six feet, measured along the pipe run, from the toilet to its vent.

A client, accepting the do-it-yourself-book's information, set his water-closet connection in concrete, seven feet from its vent, by floor measurement. When it was inspected much later, as a result of a vindictive neighbor's snitching that a permit hadn't been taken on the work, the client had to hire me to break out the concrete with a jackhammer, and re-install the toilet connection to code. The cost: over $1,000.

His modest do-it-yourself savings were gone, and he had to pay out a bundle besides, including the penalty for not taking a permit in the first place.

If he had installed the toilet to code distance, chances are that this man could have gotten by with a small penalty on the regular permit fee, since the inspector could easily see from his other, more visible work that he had been very careful and thorough in his craftsmanship.

Plumbing for Dummies adheres strictly to the *UPC*, but it is written in plain, down-to-earth language, tinged with a bit of humor and some illuminating anecdotes taken from my own wide-ranging experience as a journeyman and master plumber.

The subjects covered span simple maintenance and repairs of home plumbing fixtures, connections, pipes, and systems. I also explain construction of complete new copper pipelines and ABS or PVC drainage networks, as replacements for corroded, leaking, and dangerous existing installations. Gas piping is treated from a trouble-shooting standpoint, to assist in maintenance and repair. Plumbing fixture installations are detailed.

This book is designed to talk you through any kind of home plumbing chore you may take on. It is, however, important that you read the whole book through before you undertake any project. Plumbing is like a jig-saw puzzle—every part fits precisely to form the complete picture. You may pick up a handy hint in the section on construction which could make a repair more efficient, or easier to accomplish. On the other hand, something you may read in the maintenance or repair sections could make some fine points of sewer construction a bit clearer.

Thus far, no book on plumbing that I've read, including the *Uniform Plumbing Code*, gives the kind of broad information and helpful hints offered in this very basic manual. It contains such useful ideas as how to make solder flow uphill into a vertical copper fitting, and how to insure that your plastic pipe does not leech noxious particles into your drinking water. It explains how to make a good copper connection on pipe which would normally resist heating due to slight water drainage, by using ordinary white supermarket bread as a temporary dam to hold the water back while soldering. If, when, and how to take a permit on a plumbing job are discussed in very practical terms. Holly Evans Sammons's illustrations are clear and to the point.

I recommend that you obtain the *Uniform Plumbing Code* from the International Association of Plumbing and Mechanical Officials, 5032 Alhambra Avenue, Los Angeles, CA 90032, or from the National Plumbing Information Service, Post Office Box 3455, San Rafael, CA 94912-3455, which is *the* resource for information on current plumbing and alternative-energy systems and products. The *UPC* costs about $20 per copy.

Tom Higham, executive director of IAPMO, and his co-workers, have been generous contributors to this work, and have reviewed the entire manuscript, to insure its conformity to code.

NAPIS has assisted in the areas of new materials and plumbing techniques, and has contributed the majority of the technical solar data.

This is the right book for fearful amateur builders who dread the prospect of turning their beautiful, expensive houses into two-story swimming pools.

SECTION 1

Useful General Information

In the fifties, when I first started plumbing, as an apprentice, I was scared to try anything on my own. My journeymen straw bosses had impressed upon me that I had to be initiated into the trade by them, slowly, as befitted the normal stupidity of young people just getting started. I had to watch, listen, keep quiet, and work very carefully for four long years. Only then could I be trusted to go out into the great plumbers' beyond and pull wrenches on my own, for some of the best wages and benefits around in those days.

Steel and cast-iron pipe were the mainstays of the business, with ponderous threading and cutting equipment, molten lead seals, and heavy wrenches and machines for nearly every task.

The plumber was king in those days, because the average person couldn't afford to own the equipment and tools required to do even the simplest replacement or repair in the home.

Back then, master plumbers invoiced their services in the range of $10 to $15 per hour. Today, the spread is $25 to $55 per hour.

A person wonders about those prices in light of the fact that materials have been approved in the codes which make it possible for an average person, with average skills, to do as good a job as a competent plumber, at a fraction of the cost.

The only things left in the plumber's bag of tricks that may still frighten a weekend mechanic are the plumbing codes themselves—which imply that your work may be inspected by an official agency—and your own built-in fear of making a fatal mistake which could turn your home into a swamp either right away or sometime down the road.

The moment you cut into a closet wall or chop out a ceiling section to expose a leaking pipe or fixture, you have cast the die, and there's no going back.

I hope that after you've read this, and followed some simple instructions, you'll chop and cut away in the knowledge that you really can't do much wrong, if you use common sense.

First of all, I base everything I say on the *UPC*, as modified by some familiar regional code variations. To that, I apply over twenty years of experience as a professional plumber.

At this point, I caution you to find out if your community comes under the *UPC*, by calling your town's building department and asking, straight out. If your city has its own code or has adopted another state or regional code, then you'll have to make a visit to the building department and read through its plumbing code before you start your project.

Determine whether or not the local code allows ABS or PVC plastic pipe for sewage and drainage. Are there any limitations on the use of ABS or PVC plastic pipe in residences having more than two stories? The 1979 *UPC* limits the use of ABS and PVC to two-story homes. However, some towns take exception to that provision and permit its use in three-story dwellings. In the local town that permits the use of plastic in three-story homes, I've been told that, thus far, the experience has been satisfactory.

(If you are faced with having to install cast iron in your home to meet the code, have a good plumber deal with the iron pipe. Take my word for it, it's too much heartache for the average weekend builder. If you're made of sterner stuff, however, my suggestion is the use of "no-hub," service-weight, cast-iron soil pipe. Better tune up your muscles. Iron sewers are heavy as hell.)

In regard to the plastic pipe: ABS are the initials of the plastic formula in this pipe. It is black in color, and is joined with black solvent cement.

Some jurisdictions forbid the use of any type of plastic pipe in any kind of construction, either residential or commercial. The only drainage alternatives are cast iron and copper DWV (Drainage, Waste, and Vent).

Right here I'd like to pass on a kind of "plumbers' rule." If you can't use plastic, use copper. Copper is expensive, but much easier to install than galvanized steel.

Most plumbers will charge more for installing steel vents and drains than copper, because the labor costs are much higher for steel. Today, labor costs are higher, proportionately, than materials, no matter how costly they are. At one time, a good rule of thumb was "a dollar for labor and a dollar for materials." Now, it's about three dollars for labor against one for materials.

That's a clue for the do-it-yourselfer. Since you are paying very little or nothing for your labor, it is important to use materials that are relatively simple and easy to install. This is the best basis you'll ever find for using copper pipe.

The next item to consider, when you're looking at the local plumbing code, is whether or not you can use type M copper pipe for cold and hot water services, both inside and outside of your house. When type M is outside of your house, it usually must be buried in the ground. Where copper pipe is exposed, above ground and outside of a building, it is often required to be type L.

Type M has the thinnest walls, and is, therefore, the cheapest material to use for these kinds of jobs. The *UPC* permits it, but some jurisdictions are still holding out for type L copper everywhere. It costs a lot more money, and I don't think that such a hard-nosed attitude is in the public interest, or even realistic. It's good for the copper companies and makes plumbers' mark-ups even heftier, but I've yet to see any advantage to the heavier walls of types K and L copper tubing in residential plumbing (commercial construction is a different story).

Another product about which to inquire is PVC plastic pipe for buried, exterior cold-water installations, such as your home's main water supply, lawn sprinklers, or swimming pool circulation plumbing. Again, PVC are the initials of the pipe material formula. It is white, blue, or gray in color, and is joined by a thick, clear, gray, or milky solvent cement. (Cement color doesn't matter, as long as it is specifically designed to solvent-weld PVC pipe.)

Remember to use schedule 40 PVC for house water mains and all outside piping, which must hold the full line pressure of your water system. Thinner-walled class 125 and 200 PVC pipes may not be used to hold the full, unrelieved line pressure of your home's water system. The code says nothing about them, but they are most often used for lawn sprinkler systems, *on the sprinkler side of the valve,* where the pressure is relieved during the operation of the system. Swimming pool and hot-tub recirculation systems are also often installed in this lighter-wall PVC pipe, and it works well. Be careful, however, to check if the local code *prohibits* the use of lighter-wall PVC, in which case you're stuck with the more expensive schedule 40 pipe.

There has been some talk about this PVC pipe leeching chemicals into drinking water, and that this action can affect your health. The industry has made studies which contradict that statement—that's natural; they're in the business of selling pipe.

However, from a practical point of view, I'd like to offer my observations as a plumber who has installed every kind of pipe, tube, and conduit you can imagine. First, there is indeed some contamination in PVC pipe, immediately after it has been installed and filled with water. The problem increases when water is allowed to stand still in pipe for a few days, especially if it has lain in an uncovered trench in the hot sun. To help solve this problem, it has always been my policy to flush out newly installed pipe of any kind, especially PVC, to carry away all congealed bits of solvent cement, flakes of cuttings from sawing the pipe to length, and simple dirt and insects that may have gotten into it during installation.

Always remember to flush out your pipes, with a hose bibb or faucet nearest the delivery end of your new pipeline. Do it about one hour after you've solvent-welded the last fitting into the line. If the line is 1" or larger, three hours would be a better waiting period; you must give the cement time to harden. After you're sure that your line is tight and sound, and can see no leaks at the joints, cover all open trenches. Then flush the line once more, about an hour after the first flush, and test. Let the water run at the tap for about thirty seconds before you take your first drink.

It has been my observation that, once a new line has been thoroughly flushed, and goes into regular service, the amount of leeched chemical is so infinitesimal that it is completely undetectable, even by the most sensitive person. Copper and iron oxides, and other impurities leech out of copper joints and steel pipe in very small quantities. There are thousands of miles of asbestos-cement pipe (ACP) water mains in America, carrying potable water everywhere, and you know what asbestos does. So, what the hell, if you want good plumbing, use what the pros recommend and use, ABS and PVC plastic pipe, and copper water tube.

HELPFUL HINT: Almost all threads on pipes, fittings, and tools in plumbing tighten clockwise and loosen counterclockwise. Valves close clockwise. There are almost no exceptions to the rule in English-speaking countries. Notable exceptions are the fittings on some oxyacetylene tank fittings and valves. Some sink mixing-faucets operate right and left.

Now comes the big question. Should you take out a permit on your job? For the most part, that question applies to most jurisdictions in the United States, although there are rural areas in other English-speaking countries where the idea of permits for owner-builders isn't given a thought. Anyway, the answer is yes if it's going to run $100 or more in cost, and no if it's a simple repair of less value.

Why do I say yes in the first instance? The average person has come to distrust the bureaucracy and people who poke their noses into our private business generally, but the building and plumbing inspectors are in a different class when it comes to homeowner construction. Frankly, some of these officials can be pompous, overbearing, and irritating, and a few are not as fully qualified as they should be, but no matter what, they certainly know more about how the finished job should look than you do. The plumbing inspector is your guarantee that your project will work the way you hoped it would, when you started it. If he says that you goofed on something, chances are that your screwup is a glaring one and could be a problem breeder or even dangerous. The corrections he recommends could save your home from damage. Perhaps, they could even save your life. The cost of a permit could be the best insurance policy you ever took out.

When you take out your permit, you should find out whether or not, as an owner-builder, you must do the work all by yourself. In some states, if you hire anyone to carry out the work for you, you must take out a standard-form workman's compensation policy, acceptable to the jurisdiction involved. However, if you have help from a relative or friend to whom you do not pay wages, you are usually exempt from that provision of the law.

In many areas, the last qualification for a owner-builder permit is that you not do the work in preparation for the sale of your house. Some laws state that you should not sell within one year of the completion of your work, but there have been innumerable exceptions. That part of the law is very difficult to enforce.

I guess that this sounds like a lot of trouble to go through to get permission to work on your own house, but when it comes right down to the bottom line, what a plumbing shop may charge you $3,000.00 to do, you can do for less than half, plus the cost of your permit, if it is required.

All you need are the determination, a plan, a bit of knowledge, and a few specialized tools. If you ride a desk all week, plumbing could be the best therapy (Super Bowl Sunday excepted).

IMPORTANT NOTICE TO ALL READERS OF THIS BOOK

I cannot emphasize too strongly that, in order to utilize this book properly, you must read everything that applies to the work you are about to undertake. Don't just find a paragraph that describes techniques or materials about which you have an immediate question. Get into the habit of reviewing the material before and after that paragraph. Read all you can about the new kitchen faucet you are about to install. There are helpful hints, cautions, and special comments on every task or installation technique. Often, a hint will make your first-paragraph information more valuable, thus assisting you in your job planning and execution.

As a matter of fact, if you really want to get a feel for plumbing from the professional viewpoint, read the whole book, cover to cover. I hope you'll find it entertaining as well as informative.

SECTION 2

The Tool Box

Tools are expensive, especially good tools. But when it comes right down to the nitty-gritty, get the best tools you can afford. A good rule of thumb is, where possible, buy top-line American all the way.

Considering what you'll save on the work you do for yourself, the tool investment is a one-time deal, and it probably will not be more than a small fraction of the cost of a large job.

Bargain basement tools crack, break, don't have sharp jaw edges, and are generally a nuisance to use. Screw-downs never seem to get tight, because of badly made parts. Wrenches grab and slip. Blades wear down long before they should. Propane torch valves leak or won't open. Among common hand tools, who can forget the screwdriver blade that is so soft it warps and twists at the slightest hard going, or the Phillips driver that strips out at the first turn? Or hammers that develop soft round heads after pounding a few hundred 16-penny nails, and pliers that wouldn't hold a wet noodle? Cheap tools eventually cost more than good tools, because they fail outright or wear out so quickly. If you're going to work at a pro's job, you're going to have to invest in a pro's tools, which will probably last you for the rest of your life.

Most of my tools are over twenty years old, and are still as good as they were when I first started to use them. Not a single one of them was made in Taiwan, Korea, Hong Kong, or Singapore. The few of my tools that are Japanese are relatively new, and are based upon tried-and-true American designs. The only Japanese hand tools I would use are those guaranteed under the Sears Craftsman label. Among power tools, however, Makita and several other Japanese brands are as good as anything made here.

By and large, it's a good idea to stick with names like Ridgid, Stanley, Disston, Milwaukee, Rockwell, and the heavy-duty or industrial models of Skil and Black and Decker. They've proven themselves over the years in this country.

Here are the tools you will need to get the work done, and done well:

YOUR BASIC TOOL BOX

16-oz. straight claw hammer
9″ torpedo level
assorted screwdrivers
assorted Phillips drivers
⅜″ to 1¼″ end wrenches
Stanley Wonder Bar or equal
½″ cold chisel
¾″ cold chisel
1″ wood chisel
center punch
nail set
12″ Channel Lock curved-jaw pump pliers
12″ hacksaw with several 32-teeth blades
25-ft. measuring tape
5⁄16″ nut driver
safety goggles
lead pencil
12″ pipe wrench
18″ pipe wrench
12″ crescent wrench
8″ crescent wrench
utility razor knife
half-round metal file
110-volt work light
work gloves
electric drill and bits
7″ hand-held power saw
small aerosol of WD-40
small can of *3-In-1* oil
medium can, best-quality pipe-joint compound
small can of waterless hand cleaner
large box of bandaids

Approximate cost at this writing: $425 plus tax.

MAINTENANCE TOOLS

12″-14″ basin wrench
faucet-valve seat-grinder
assorted faucet valve-stem washers
closet-auger (special snake for toilets)
Drain King drain opener— Manufactured by: G.T. Water Products, Inc., 19438 Business Center Drive, Northridge, CA 91324
Model 501 for 1″-2″ drains
Model 186 for 1½″-3″ drains
Model 750 for 3″-6″ drains
plumbing-fixture faucet-handle puller
professional force-cup (plunger-plumber's friend) with flange or apron
10-12 ft. standard spring-snake

CAUTION: *The Drain King is in violation of section 1003(e) of the UPC, which deals with direct connections between potable water piping and sewer connected wastes. However, in my years of experience with the tool, I have never had a backflow of sewage into the drinking water system of any house or building in which I've used it. According to the inventor and manufacturer of the device, Mr. George Tash, who is vitally concerned with the safety of his unit in use, he has never had an adverse report on the Drain King with respect to such backflow.*

You are cautioned to use this device with a hose-bibb backflow prevention device.

The reason I recommend Drain King will become evident in Section Three of this book. It works as no other device in certain drain unblocking situations.

However, the ultimate decision is yours. If you use Drain King, be sure that you watch it carefully during use; if you have water-failure during its use, disconnect the hose from the faucet or hose-bibb immediately; *and when your water service is restored, flush out all taps for thirty seconds to insure that you have removed any small amount of backflow which may have entered the lines.*

Approximate cost of maintenance tools at this writing: $75 plus tax.

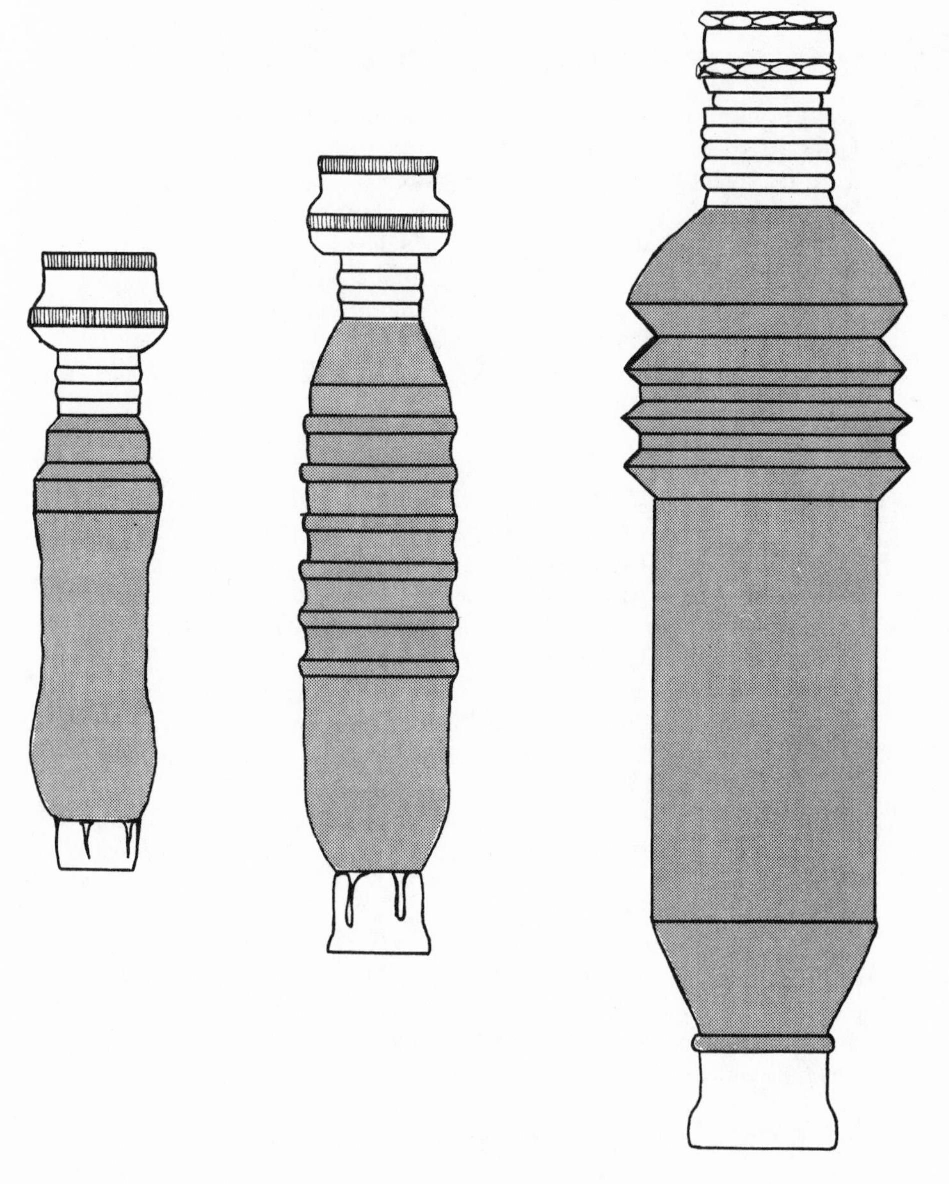

The three models of the Drain King: *left,* Model 501 (small); *center,* Model 186 (medium); *right,* Model 750 (large).

COPPER PIPE INSTALLATION TOOLS

standard propane torch, w/ tank (hand-held assembly)
1 container of AMCO-C flux
1 sleeve of medium steel wool
1/8″-3/4″ copper-pipe cutter
1-lb. 50/50 solid-core wire solder

Approximate cost at this writing: Under $45 plus tax.

PLASTIC PIPE INSTALLATION TOOLS

Sandvik special saw for plastic (If unavailable, use hacksaw)
rubber gloves
solvent-cement of the appropriate type for the pipe to be joined
self-constructed pipe-cutting stand

Approximate cost at this writing: Under $25 plus tax.

DIGGING AND PREPARATION TOOLS

cross-handle, diamond-point shovel
pick-mattock
8-12 lb. sledge hammer (long handle)
24″ wrecking bar
4-8 lb. hand sledge (optional: not calculated into cost)

Approximate cost at this writing: Under $60 plus tax.

It is understood that maintenance, copper-pipe-installation and plastic-pipe-installation tools are in addition to the general tools at the top of the list. You must have your basic tool box to do any kind of plumbing work.

Remember that even top-line tools can be bought at discount through several catalogs and nation-wide discount houses.

Shop several stores, catalogs and outlets until you find consistently lower prices for all of the tools on your list. Don't worry about the odd wrench or screwdriver that may be priced a few cents lower at some other hardware store or supply house. If the "big-ticket" items, such as the power saw, drill, and large wrenches, are fairly discounted, along with reasonable prices for the rest, buy all of your items from the same supplier. In this way you will establish a relationship for later returns of defective tools, if that should become necessary.

Remember, *anything* manufactured *anywhere* might come up defective. You will need the clout of loyal patronage to smooth the way to hassle-free replacements.

Sears will replace any of its Craftsman hand tools, without a negative vibe. I've taken back screwdrivers that have obviously been abused after two years, without a sales slip, to a Sears store at the opposite end of the country from where I originally purchased them, and have had them cheerfully replaced, without a problem of any kind. It's an unbelievable experience these days, and is probably the most enlightened promotion idea in American merchandising.

With the tool list I have proposed, you should be able to do any kind of plumbing repair and construction necessary to maintain and upgrade your home. It is obvious that your basic tool box contains items which will be useful in all types of home construction and repairs, and, in fact, most of the folks who buy this book have the majority of them already. All that must be done is to add the missing tools. For some, it will mean a very small investment.

For others, a couple of hundred will do the job.

If you have no tools, and are simply not "into" working with them, I wonder if this book is for you. Perhaps you ought to stop here, before you've soiled it, and return it to the bookshop. I'm dead serious. If you haven't had much experience with hand tools, chances are that your mechanical skills are not well-enough developed to do plumbing, which is very demanding work. Why do you think plumbers are able to charge as much as they do and get away with it? Think about it, and if you feel that you'll be able to cut it, go on reading.

Incidentally, the best mechanic I ever apprenticed was a woman who stood 5′ 2″ tall and weighed a little over 100 pounds soaking wet, which she was much of the time. She outworked and outthought most of my 160-pound male apprentices and several heavyweight journeymen. As a point of interest, Sal made journeyman, later took her master plumber's license, opened a shop in San Jose, California, and was sailing right along, until she met a happy-go-lucky Irish plumber from Calistoga. She got married and I never heard from her again. But I'll bet you their home has the best plumbing in Napa County.

To wind up this section of the book, I'll make some additional suggestions:

It's a good idea to keep your tools in well-made steel tool boxes which have provision for locking. All power tools should be kept in separate carrying cases.

After you have worked in or around water or moisture, make sure that you dry and oil your tools to keep them from rusting. Always put them away, in their boxes, after work is finished for the day.

Periodically, lay them all out and give them a good cleaning with kerosene or turpentine. Then oil them lightly. Minor rusting can be removed easily by washing in kerosene, drying, and then giving the rust spots remaining a brisk rub with steel wool.

Take good care of your tools, and they'll take good care of you.

SECTION 3

How to Maintain and Repair What You Have

What you maintain, you will seldom repair or replace. Maintenance usually means keeping your drains clear, stopping faucet leaks, and making sure that your pipes are sound and dripless.

Preventive maintenance is the best kind. It is simple and direct. It usually works, and it can save you pots of money over the years. In fact, one of my suggestions will probably save you the price of this book, several times a year.

The most common causes of stopped-up drains are hair buildup, paper, women's sanitary materials, grease buildup, male prophylactics, kids' toys and sagging drainage piping (bananas), in that order.

Every one of them can be prevented.

Hair buildup is very common in tubs, showers, and lavatories (bathroom sinks). In some homes it also occurs in the kitchen sink, but not so often. Women are frequently accused of causing this situation, but believe me when I say that a good amount of shower-drain buildup is man-caused. The easiest and best way to eliminate this problem is to buy small drain strainers, snap them into place, and clean them out, after every shower or hairwashing. Get into the habit of cleaning out the strainers, or else very soon you will find them being set aside, kicked out of place, or otherwise bypassed by impatient family members who are anxious to get the slowed-up drain going, after the strainer fills up and blocks the water flow. They don't think of cleaning the strainer out—that's too much trouble and it's yukky. Okay, then, let it go, and if you can't use the drain-cleaning methods that I give in this book, call the Roto-Rooter man and pay over $30 for the privilege.

If your drains are not properly sized for standard strainers, of if there is any other reason why you can't use a strainer, there is another method of preventive maintenance:

There are any number of commercial solutions and compounds that can be poured into drains to clear them of blockages, but it's an exasperating fact that, when drains become totally blocked, quite often these drain unblockers fail miserably. The reasons are many and complicated, but it's enough to say that sometimes chemicals react to soap, grease, human hair, and some kinds of plastics in very strange ways. Under certain circumstances, using an unblocker can make the situation even worse. An example: If your kitchen sink is stopped up with grease and you use one of the lye-based unblockers (sodium hydroxide), the lye will make soap when it comes in contact with the grease. A grungy, rather hard cake of soap will exist where once there was softer grease.

For effective removal of hair buildup, after you've removed all visible waste material from the drain from the top, twice or three times a week use one of these unblocking compounds or solutions. They work best when there is a minor buildup in the pipes. Doing this on a regular basis will keep the drains from building up to a major blockage. Make sure that you get a type of unblocking compound or solution whose label plainly states that it dissolves hair and is effective on grease.

To sum up, when you have hair blockage problems, try to turn the situation around through preventive maintenance, using strainers if you can, especially in fixtures in which the family washes hair, and certainly in showers and bathtubs, which collect body hair as well as head hair. For those of you who have pop-up drain stoppers, strainers are out for most types of bathroom lavatories (sinks) and tubs. In those cases then, the only answer is regular applications of drain-unblocking solutions or compounds.

In the case of blockages by toys, preventing the problem is as simple as can be. Most toys are either sailed in tubs or playfully dropped into toilets. (I can still remember how much fun that was, and how sore my butt got when my mom or dad found me doing my thing.) Toilet toy-disposal can be prevented by tying up the flush-handle on the water closet's tank, so that it can only be flushed by an adult who can read the instructions which you post on the wall behind the tank. There are lots of ways to do this, including forming a piece of flat metal band to catch under the back of the tank lid, ride over the tank top, and wedge under the flush-handle. A spring-loaded string or wire attached to the handle is another way, but it's not as brat-proof as the flat-band method.

There's another way, but it's a bit more trouble. After every flush, you turn off the water at the valve beneath the toilet tank. The principle of all three methods is that if the toilet can't be flushed by the infant-culprit, the toys and debris will remain in sight and can be removed (ugh) by daddy, mommy or good-natured guest. Errors in judgment, a heavy night on the town with spontaneous upchucking at home, sitting down without looking, other kids, and a zillion other things, can make any one of these methods useless and cause for serious cussing and grousing. If you have two water closets (toilets) in the house, the problem is solved. One is the child's exclusively (you hope), and one is only for grownups. This is a challenge to a child, so keep eyes peeled and plungers at the ready. Keyed door locks may be indicated for toilet rooms, if all else fails. Make sure that you put the keys in plain sight and beyond the reach of sir or miss bratnick. A small bell to be rung near the appropriate door could be used to gain admittance by your junior toilet villain.

Then the day comes, when everything you've done, including putting the potty in a vault, comes to naught. The can is completely blocked by plastic alligators, cable cars, Greek busts, and a fistful of toy balloons. These will probably have to be removed mechanically.

The only way you'll ever get paper, female sanitary products, condoms and disposable-diaper plastic liners out of drains is by hand, or with special tools designed specifically for that purpose.

Important: Never throw used condoms, tampons, tampon-inserters, sanitary napkins, disposable diapers, paper towels or napkins, and Scotch-type scrubbers in the toilet.

If a condom goes into the toilet-trap in a certain way, it will fill with water, like a balloon, and make a permanent block in the water way. Tampons are designed to accept tremendous amounts of moisture, and can swell to unbelievable size in water. Tampon-inserters are often stiff plastic tubes which can become lodged in the trap system. One inserter won't stop much, but the combination of a tampon and the inserter, plus a bit of toilet paper winding around both of them, can cause a serious blockage and an overflowing toilet.

Okay, so your toilet is a disaster, full of toys and other debris, and you want to do a complete job of clearing it. There is a special sequence to be followed, ranging from simple to drastic procedures.

CLEARING WATER-CLOSET (TOILET) BLOCKAGES

The first thing to do, of course, is remove the visible debris all the way down into the sump of the toilet bowl. This can be a most distasteful job, so long rubber gloves are recommended to keep the soiling to a minimum. However, in fairness, I must observe that using bare hands is more effective. You can feel more of what is there, especially small objects. When you've completed this disgusting task, naturally wash your hands thoroughly, not only for cleanliness but for the simple reason that you don't want to soil the equipment you must use in the clearing process.

Now, with all the toys and other hard objects out of the bowl and sump, get your plunger and pull out the apron or flange. The flange type is recommended in Section Two (The Tool Box). This flange is designed especially for clearing water closets. At this point, place the plunger in the bowl and plunge away to get the water level down as far as it will go. This may clear the blockage immediately, but most of the time it won't.

With the water level down, lift the plunger away from contact with the bowl and flush the toilet. As soon as the water has risen above the sump about 6", quickly place the plunger into the sump and pump as rapidly and vigorously as you can, lifting the plunger away from the sump opening before the water rises to the toilet rim. At this point, if there is no solidly emplaced piece of material in the water way, the toilet should empty with a rush, cleaning out the remaining blockage.

If the toilet overflows, and does not drain rapidly, it is obvious that you have a serious obstruction in the toilet's internal trap system.

Plunger in Proper Position for Clearing Toilet Blockage

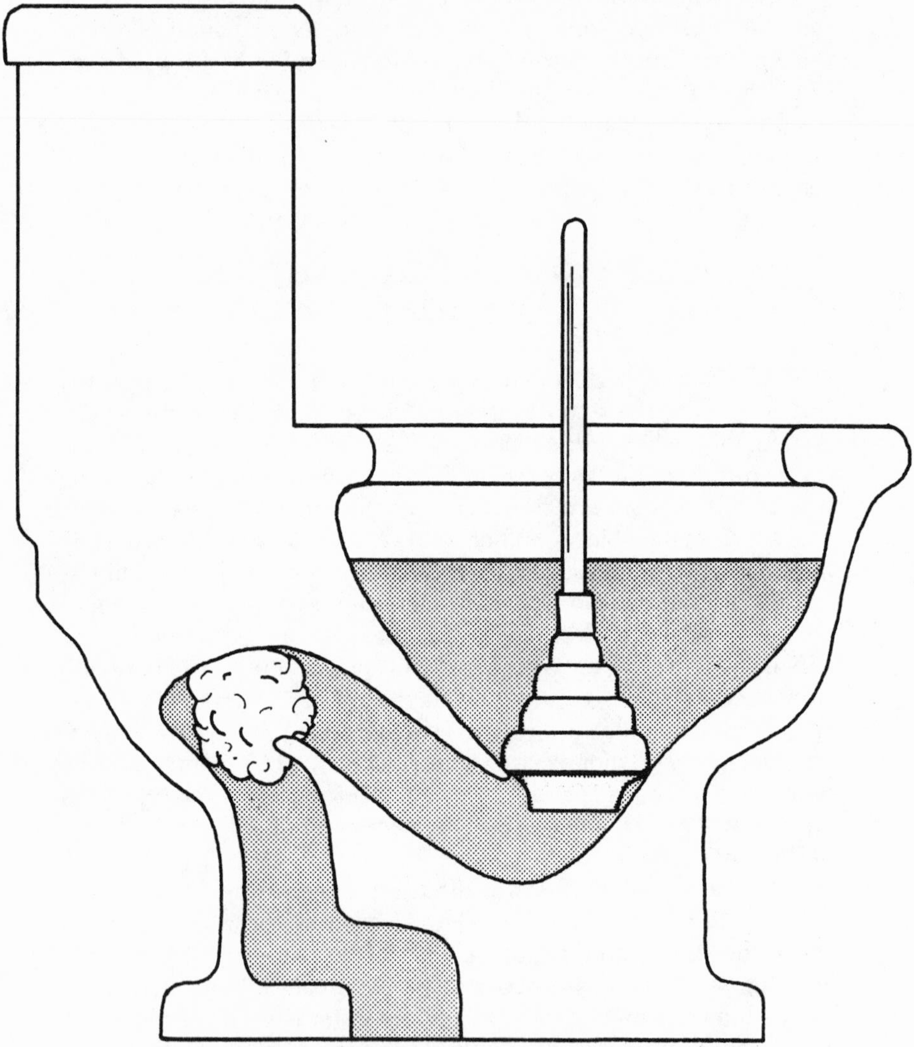

Plunger in toilet with flange extended, water-level high in bowl, and nondescript stoppage in toilet trap.

The only thing to be done now is to pump your plunger sharply until you have again lowered the water level to the sump. With the input of so much fresh tank water during your plunging procedure, the foul water will have been diluted to the place where you probably won't object to putting your hands into it. You must now further drain the toilet with a cup or some other small container which will fit into the sump. When you have the water down as far as it will go, get a small mirror and lay it in the sump, angled in such a way as will permit a look back into the upper trap. It is helpful if you get a flashlight and direct it into the angled mirror, to light up the interior of the sump tunnel or channel. You should then be able to see any obstruction in the sump entrance. If you find one or more, you should be able to fish them out with a wire coat hanger, bent into a shape which would fit behind the obstructions and allow you to pull them into the sump. If that won't work, then use your closet auger (see Maintenance Tools, page 9). Place the plastic or rubber tip of the closet auger down into the sump, with the spring-tip facing the obstruction. Make sure that you pull the spring all the way out of the metal-tube holder, and that the tip of the auger is nested against the plastic tip of the tube. Now, push the spring down to force the auger tip into the sump, while turning the crank handle. You may find it awkward to turn the crank, with so much spring out of the tube. The object is to force the tip forward from the plastic tip, into the throat of the sump and, from there, into the waterway and trap of the toilet. Often, just a sharp push downward will slide the auger tip right into the throat, and you can begin turning the crank, pushing down continuously. Watch out for the possibility of the spring doubling over on itself and just winding into a kind of knot. If that happens, don't worry. You'll still be able to get the auger tip out again, but you will have done little effective work with the tool. Just repeat the procedure until you get the hang of it—pulling the spring out of the tube, setting the head of the auger in the plastic tip, placing the auger with the tip and spring-head down into the sump, facing the obstruction, pushing down sharply with the spring crank-handle to thrust the head into the throat of the sump (entrance to the internal waterway and trap), and then cranking the auger to snag any objects in the trap system. Now, you withdraw the auger. You should bring up some debris, including the always present wads of toilet paper. Perhaps you'll luck out and snag the main obstruction,

The Mirror-and-Flashlight Toilet-Sump-Inspection Method

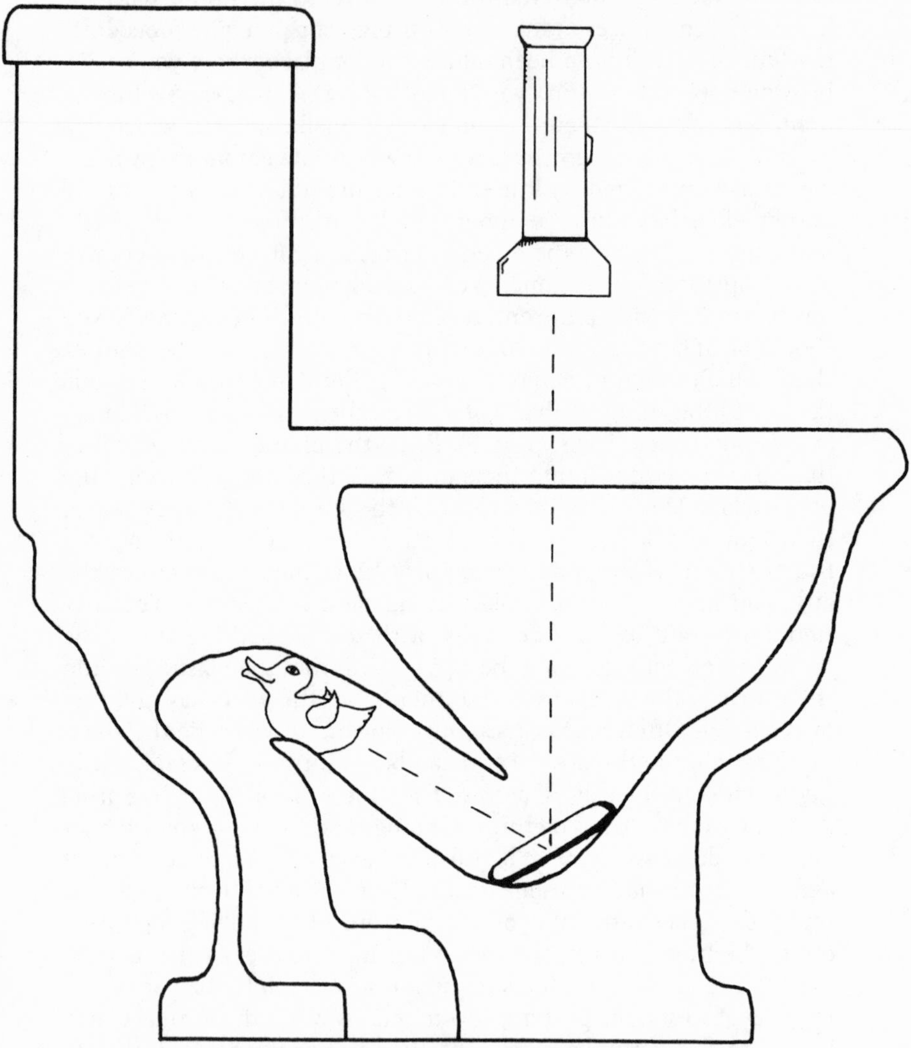

Side section of toilet with "ducky" obstruction in trap, and flashlight shining into a mirror in the sump, illuminating the obstruction.

Removing an Obstruction from a Toilet, with a Wire Coat Hanger

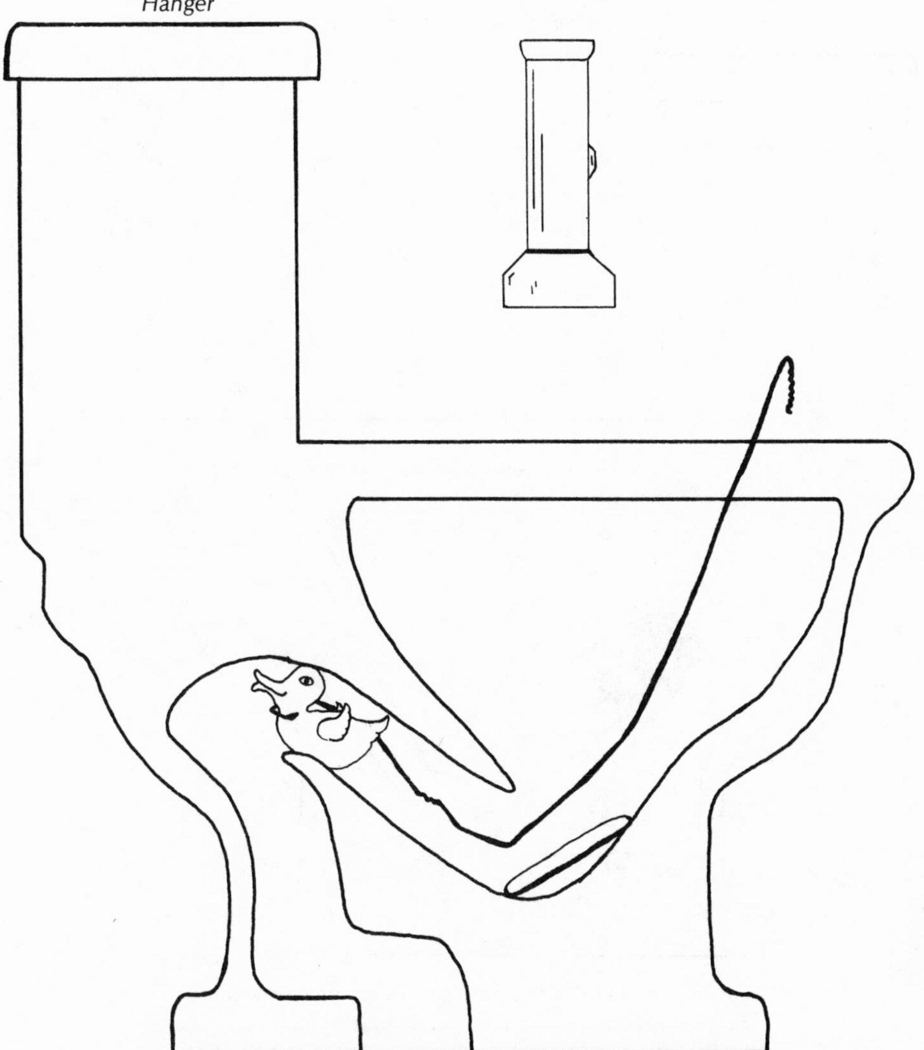

Side section of toilet with "ducky" obstruction in trap, and opened-out wire coat hanger inserted to grasp obstruction.

Closet Auger Removing Obstruction in Toilet

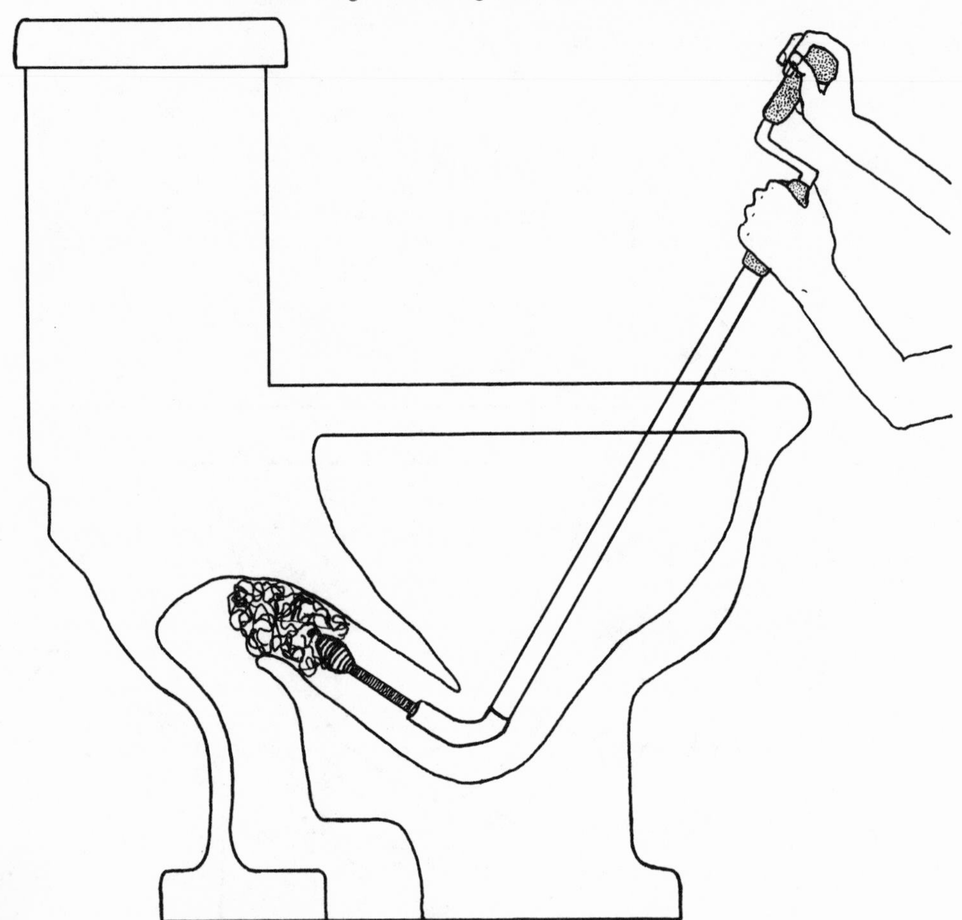

Side section of toilet with obstruction being snagged by an L-shaped closet auger.

or several pieces of plastic which you may have shattered with the auger head.

The next move is to repeat the plunging technique which I have explained in detail. Follow the same instructions, only this time, if you don't get an immediate rush of water out of the bowl, plunge it once more with all your energy. This should be enough to get the job done, but if it isn't, give the closet auger another try with the bowl up to its normal water level. This time, push the spring as far as it will go before pulling it out. More debris will probably come into the bowl water, which you should again remove by hand. Now, try the plunger sequence again. If you fail to clear the blockage, you must go to the next step.

DISMOUNTING THE TOILET

Removing the toilet from its mounting ring (closet flange) should normally be done by a very strong person, or better yet, by two people working together.

Before you even start, go to your plumbing or building materials supply store and buy a new water closet wax ring *with plastic flange* (they come with and without), and a new set of closet bolts.

HELPFUL HINTS: Most do-it-yourself-type stores have these bolts in cello-packs. Closet bolts come in two types. One type looks like a wood screw on one end and has ordinary bolt threads on the other. This one is designed to screw right into a wood floor, and then, when down and in place, you put the toilet in place over both screws, and bolt up with the cap-nuts and washers. The other type is all machine threads, with a flat head on one end. This type is designed to slide into the closet flange slots. When they are in place, again you position the toilet over the closet flange and let the bolts pass through the holes in the base. They are then bolted in the same fashion as the wood-screw type. This last type of closet bolt has a tendency to wobble and get out of line when you're putting the toilet back in place, so I usually use a bit of the old wax from beneath the toilet to secure the loose bolts in the upright position. It's a first-rate method of managing this frustrating problem.

In any case, when you buy your new set of closet bolts, most stores have both types packed in one rack-pack. Of course they have each type in individual packages, as well. My suggestion is that you

One Method of Removing a Toilet from Its Flange

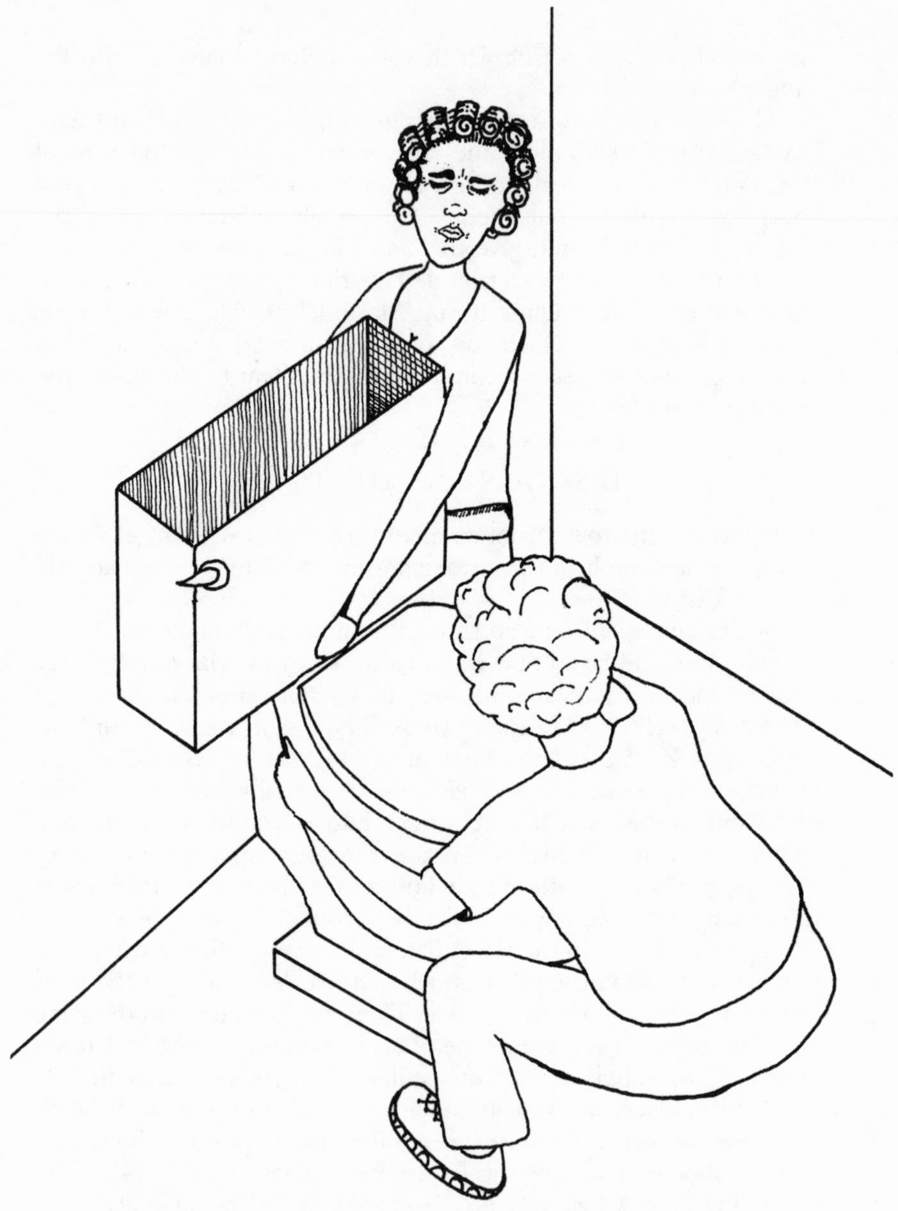

A woman in curlers and another person lifting a toilet from the floor.

buy the double pack, containing both types of bolts. If you have a wooden floor and it is sound beneath the bolt slots of the closet flange, by all means use the wood-screw type of closet bolt. You won't have to worry about packing these with bowl wax to keep them upright while you replace the toilet on its flange, and these are often more positive fasteners, especially if your closet flange is made out of ABS plastic. An additional advantage is the fact that often you needn't replace this type when you have to pull the toilet. They are so solidly fastened into the floor that the nuts can be unscrewed, even if they're rusted.

Don't worry, however, if you don't have a solid base for the wood-screw type. Most toilets are fastened with the tee-head bolt that slides into the flange slots, and they work very well.

The reason you must buy the new ring and bolts in the first place is because 90 percent of the time you must discard the closet bolts now holding the toilet in place—the act of removing the old bolts will usually ruin them. In many cases you will have to hacksaw the old ones off. *You must always replace the wax ring.*

At present prices, the new parts should cost $7.00 or less.

Now, you're ready to clear your toilet.

First, remove the tank lid and place it in a safe place, outside of your work area. Lids are almost impossible to replace if they become cracked or broken.

Next, shut off the water supply to the toilet tank, by tightening the angle valve (turn clockwise, as you look at it). Take it to the fully closed position. You may have to help it along with a pair of pliers. It is important to remember that, whenever you are opening or closing the small supply valves serving sinks, lavatories, and water closets, never overtighten them with wrenches or pliers. They are not designed to take much pressure on their handles and can easily be overtightened and break. Tighten only far enough to shut off the water completely. The way to be sure is to flush the toilet, and listen to the ballcock (input water regulator with the arm and ball, or similar valve). If you still hear the water hissing through the line to the tank, the shut-off angle valve underneath the tank should be tightened more and more, until there is no more hiss or bubbling in the toilet tank. Then you know it is completely shut off.

In the event the valve is so old and worn that it will no longer shut

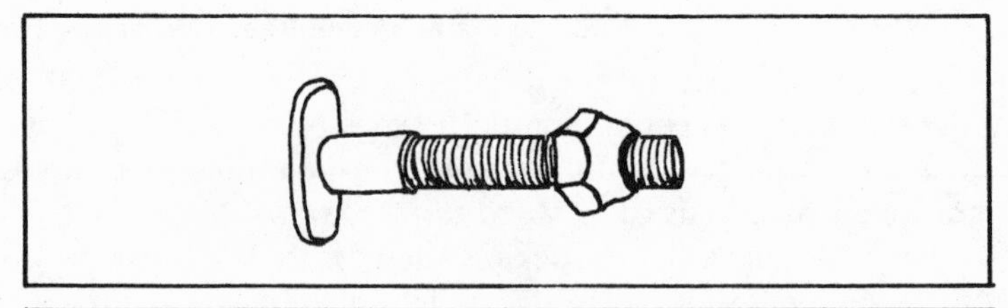

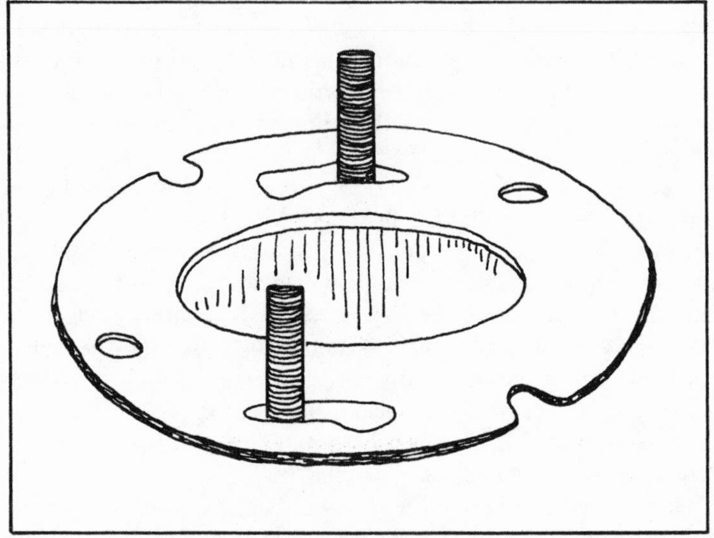

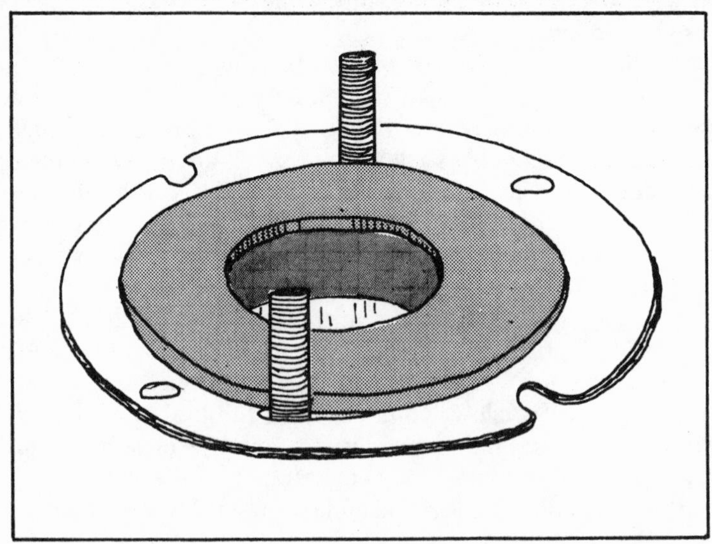

Top: closet bolt with nut; *middle:* closet flange with closet bolts in place; *bottom:* closet flange with closet bolts and bowl wax in place.

off completely, you must shut off the main house water supply either at the main valve or at the meter.

Now, flush the toilet again to remove as much of the tank water as possible. It is best to remove all of the water from the bowl and sump. Again, pump with your plunger until you've gotten the water down to the sump or lower. Then dip the rest out with the cup or a tin can. You can usually sop the rest of the water out of the toilet tank and sump with a large sponge, or with other absorbent materials. You do this so that, when you take the toilet off its ring, a minimum of water will escape when you move it. As you will see then, no matter what you do, or how you do it, you are going to have water from the toilet, and it will probably be somewhat more polluted than the bowl water which you have so carefully dipped out.

Because of this problem, it is suggested that, if the toilet is going to be placed upon rugs or other decorative surfaces while it is being worked on, you should lay it on a thick mat of newspapers, at least six feet square and about ¾″ to 1″ thick. Have a mop handy, and be prepared to control from a few ounces to as much as a quart of water. Better yet, if the lawn is close to the toilet's location, by all means do your work out in your yard. But, even there, place the toilet on a bed of newspapers. It makes the project a bit easier to manage.

With the supply water completely shut down, and your work area prepared, you are ready to remove the toilet.

First, loosen the top nut on the water supply tube, right up under the tank. Remember: Turn the nut counterclockwise, as you look at it. Either a pair of pump pliers or your 8″ crescent wrench should do the job quite handily. If you are in really tight, and can't get the handles working to turn the connection, use the basin wrench. The basin wrench is the strange tool that looks like a 12″ steel rod with a tee handle and two half-moon-shaped jaws on a swivel at the other end. A little common sense and practice will show you how to operate it.

If the basin wrench won't work, then there is only one other way to go. Loosen the small nut located on the angle valve that you have closed. The nut will loosen if turned counterclockwise, as you look at it. Usually, the valve and nut are chrome-plated. Use an end wrench on the nut, to avoid scratching and distorting it.

I prefer that you unloosen the upper nut near the tank because,

Basin Wrench in Action

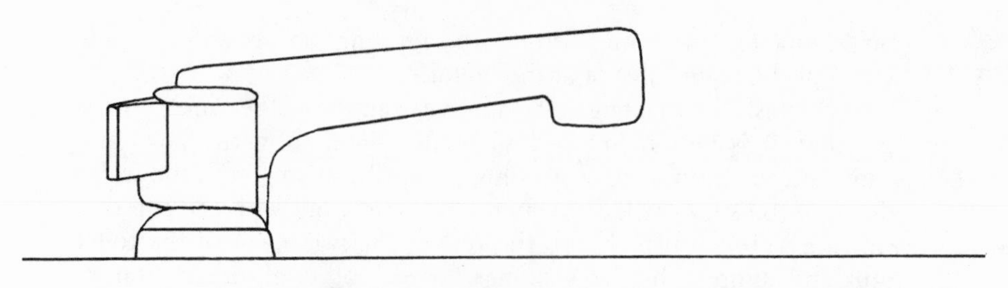

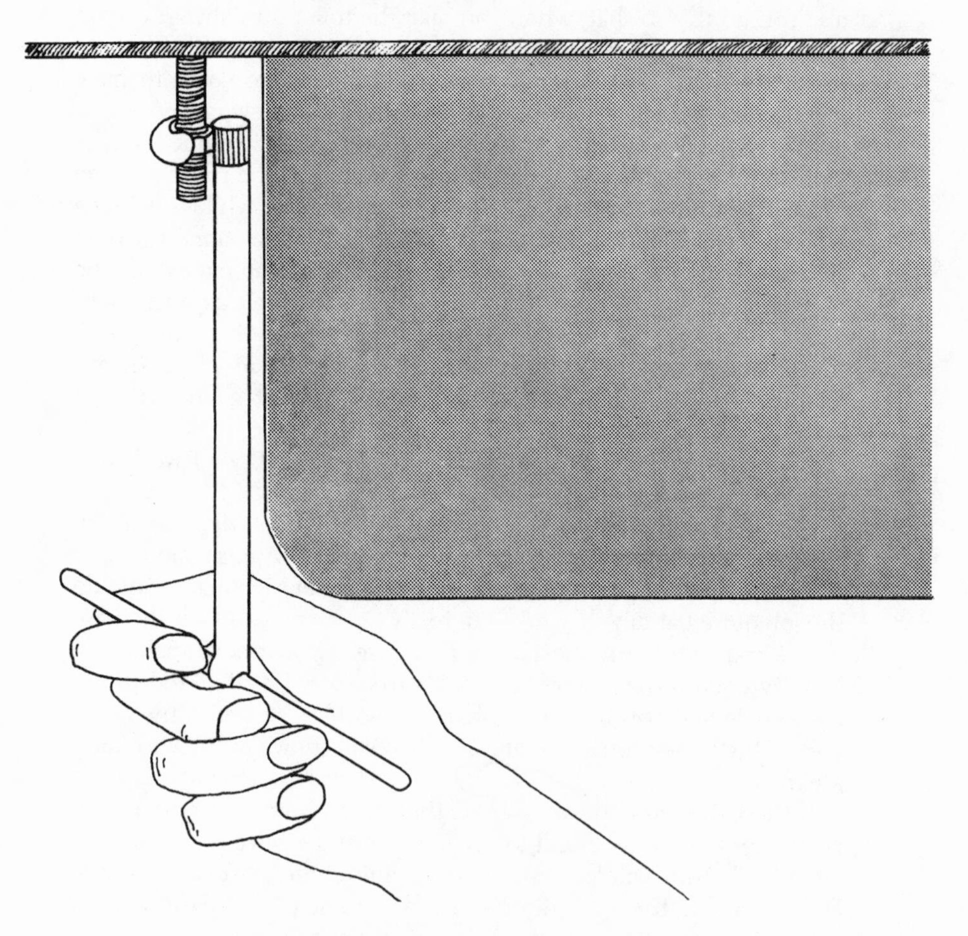

Right hand operating knurled-tipped T-handled rod, from which projects a curved jaw, grasping the nut on a faucet barrel, with the cross-section of a sink and faucet.

when you disconnect the water at the valve, you leave the very delicate chrome water-supply tubing attached to the toilet, and it can become damaged or bent beyond repair while you are moving and working on the toilet. So, if you have to disconnect at the valve, be extra careful of that tubing when you move the toilet.

With the water disconnected, the next step is to remove the bolts at the base of the toilet. Usually, they are under fancy caps which are designed to hide the bolts and nuts themselves. All you need to do is pull up sharply on the caps and they should snap away. If they don't, use whatever force it takes to remove them, even if you break them to pieces. Later, you can go by your supply house and get a pair of plastic caps, which are cheap and very easy to install.

Try, at first, to unscrew the nuts with a small end wrench. Most of the time it won't be easy, especially if the toilet has been in place for ten years or longer. Quite often the bolts will turn with the nuts. If there is a bit of the bolt showing above the nut, you can try to hold the tip with a pair of pliers, while you try to start the nut unscrewing. Sometimes a bit of Liquid Wrench or WD-40 helps to free an obstinate closet-bolt cap-nut.

If everything you try fails to budge the nut, or continues to turn the closet-bolt itself, there is nothing left to do but to hacksaw the bolt and nut off. Older bolts and nuts were made of brass. They'll hack out like a hot knife going through butter. The brass-plated steel ones are a bitch. Don't worry about your saw blade making black marks on your toilet. The china is harder than the blade and will wipe off with a damp cloth and a bit of kitchen cleanser or scouring powder. If you are working in tight quarters, take the blade out of your hacksaw and wrap one end in rags. Carefully, using the rag-wrapped end as a handle and sawing in short strokes, saw the closet bolt below the nut, if it is possible. If it isn't, saw as low on the nut as you can, and when it's completely cut through, put a putty knife or other broad, strong blade under the nut and lift up sharply. That should be sufficient force to pop the rest of the nut off the bolt.

Now, go to the other bolt and repeat the sequence, first trying to unscrew the nut, and then going to the more difficult job of sawing.

Once both bolts are free, there is nothing holding the toilet in place except the natural stickiness of the wax ring which seals it to the drainage system.

The best way to get it loose is to tip the toilet forward, by taking hold of the tank, drawing it toward you, lifting at the same time to

Angle Stop with Supply Tubing and Fittings

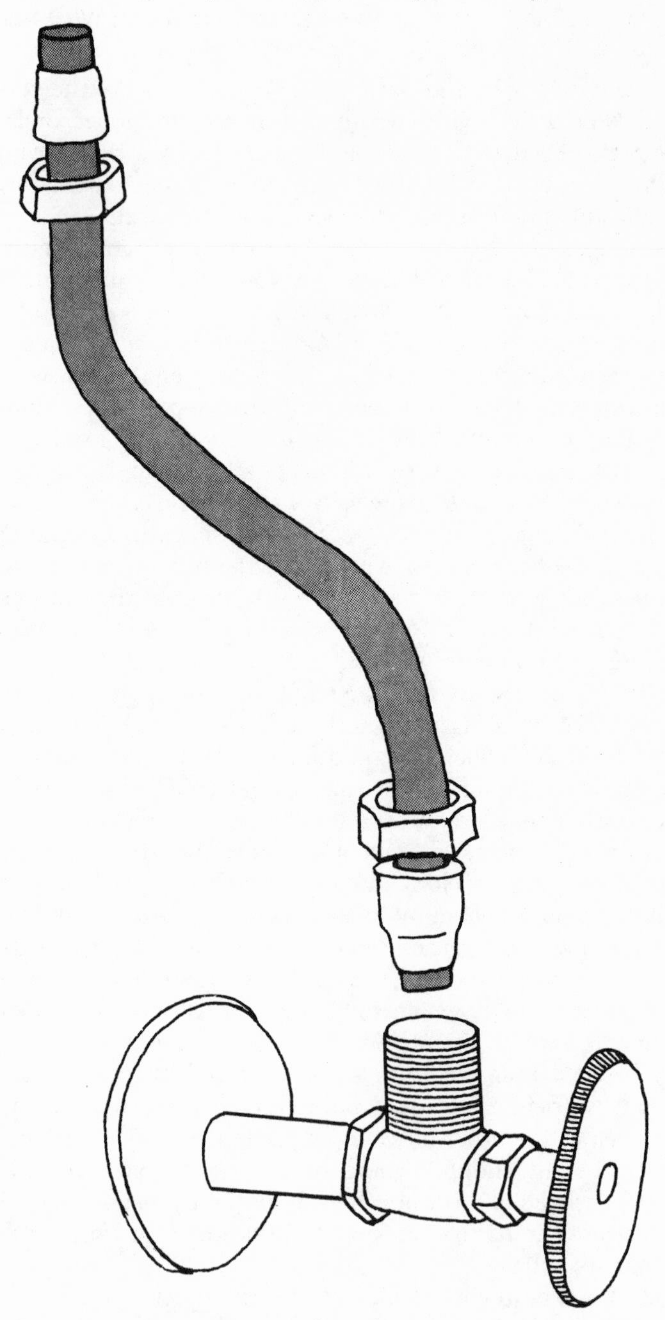

Angle stop coming out of a wall, showing separated supply-tubing ferrule bands and two nuts.

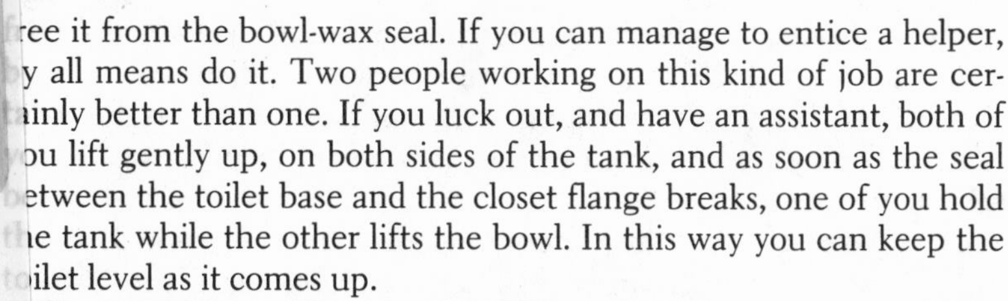

free it from the bowl-wax seal. If you can manage to entice a helper, by all means do it. Two people working on this kind of job are certainly better than one. If you luck out, and have an assistant, both of you lift gently up, on both sides of the tank, and as soon as the seal between the toilet base and the closet flange breaks, one of you hold the tank while the other lifts the bowl. In this way you can keep the toilet level as it comes up.

HELPFUL HINT: The more level the toilet is as it is moved to the place where you're going to work on it, the less water will spill out of the waterway and trap system in the base of the toilet. When you get it to your newspaper-prepared work area, let it down *very* gently, keeping it upright, while one or the other of you gets a large shallow container of any kind, preferably a shallow baking pan or a foil throwaway oven pan. Now lift the toilet straight up and turn it on its side up in the air, with the outlet hole in the toilet base over the shallow pan. If you have removed all of the water from the tank and sump, you should have no water spillage from either the tank or bowl, but, most likely, you'll have water coming out of the base opening. This should be caught in the pan, so that when you lay the toilet down on its side, if the base drips, it will go into the pan. Now, you're ready to clear the blockage.

At this point I should discuss the type of toilet that has the tank mounted on the wall, and a chrome-plated flush-elbow connecting it to the bowl. Working on one of those is one heck of a lot harder than taking out the close-coupled type which I've been describing. Sometimes, you can unscrew the nut that fastens the flush elbow to the bowl and, at this stage, lift the bowl up and out without much difficulty. Most of the time you'll have to take the tank off first, which means loosening the top nut of the flush elbow on the tank, as well as the bottom one on the toilet. With the flush elbow loose, you can try once more to lift the bowl, without taking the tank off the wall. If that fails, then you've got to unscrew the tank from its carrier. Often, it is just set onto the carrier lugs, and can be lifted off without unscrewing any tank bolts. There just won't be any there to unscrew. But, if you see that the tank is held in place by large brass screws or bolts, going through holes in the inside back of the tank and into the wall, get them out, get the tank off, and set it aside.

Most of what happens after that is the same for both types of toilets.

You've tilted the bowl forward. If you're by yourself, get a good grip on it, with your arm circling under the tank on the modern-type toilet, to support the tank, and lift the whole toilet off its ring. Remember that the old wax ring holds on quite tightly to the toilet base, and may cause you to lose your balance. In any case, if you have no help, just brace yourself.

Once you've gotten the water closet over to your work area, and on its side, you may continue without assistance, until you have to return the toilet to its proper place.

If you've removed the toilet to the yard, it may be a good idea to put a hose in the trap hole and flush out the polluted water before you place it on its newspaper bed. You are going to reach into the toilet trap and remove all the obstructions that you may find there.

After you have cleaned out the trap and waterway, all that is left to do is remount the toilet and bolt it up.

Before you do that, however, you *must* clean all of the old wax from the toilet base and the mounting ring (closet flange) in the bathroom. At the same time, slide out the old closet bolts and replace them with the new ones. As I mentioned before, take a bit of the old wax and put a small amount on each tee-type bolt, when you've slipped it into its mounting slot on the closet flange. Build a small mound of wax around the base of the bolt, sticking it to the flange so that it's standing right up. That will make it relatively easy to drop the toilet into place without having to worry about the bolts moving out of position.

With that done, place the new wax ring in place over the iron or plastic floor ring (closet flange). With someone to look beneath the base to guide you, drop the toilet base into place gently, with its protruding trap hole right in the center of the wax ring, and the bolts coming through the slots in the toilet base. Push down and move it a bit to get it to spread the wax underneath. If you can't get it to go all the way to the floor, just sit on the toilet and roll your hips a bit to help work the toilet bowl down. Eventually, it will rest flat on the floor, and you'll be ready to bolt up. Make sure that you place the washers between the nuts and the toilet base.

HELPFUL HINT: If you are going to put the snap-on type of plastic bolt caps in place after you've tightened the nuts, you must first attach the plastic mounting discs between the closet-bolt washer and the toilet base. Usually, the discs have instructions embossed on their surfaces, such as "This side up," etc. If they don't, put the beveled side down.

When you have tightened the nuts, and the bowl is rigidly in place, grab the bolt shank above the nut with a pair of pliers, and flex it back and forth until it snaps off flush with the nut. If there is little or no shank above the nut, of course you leave it as it is. The only reason you try to keep the bolts down near their nuts is so that you can replace the caps and seat them solidly to the toilet base. If your caps are the type which are held in place with putty, simply scrape out the old putty, refill the cap with new plumber's putty, and press it into place.

Now, reconnect your water supply, and your toilet is ready for service.

UNBLOCKING OTHER FIXTURES

All other fixtures are much easier to unblock mechanically than are toilets, but, as with the toilets, you should try other methods first.

There's a reason they call the force cup or plunger "the plumber's friend." It is a second line of offense, when preventive maintenance is not used regularly.

Whenever you use the plunger on sinks, showers, or tubs, the flange or apron is pushed into the cup, making your professional model look like the plunger you've been used to seeing in nearly every home you've visited since you were a kid. The apron is only used to clear toilet blockages.

PLUNGING KITCHEN-SINK DRAINS

To clear a kitchen sink, simply fill it with hot water to within about 2" from the top. Place the plunger cup right over the drain hole or garbage-disposer opening. Pump sharply for a few seconds and lift the cup from the drain. It should be clear and the water should drain out quickly. If it doesn't, repeat the procedure several times. If it doesn't work, repeat the procedure several more times. If you fail again, more drastic action must be taken.

Some things to check out, if your first tries fail: If you have a double-compartment sink, you have two drains. You must plug up one of the drains before you plunge the other. The best way to do that is to ball up rags into a tight fitting plug and wrap the ball in a plastic sandwich bag, plastic wrap, or the corner of a small garbage bag.

Stuff the rag-filled plastic into one of the drain holes, and plunge the other. You can use the help of another person to hold the plug in place. If there is no help available, you'll have to manage as best you can, but a good idea is to fill both compartments to the same level with water. The water on top of your "rag plug" will help to hold it

A Double-Compartment Kitchen Sink Properly Prepared for Plunging

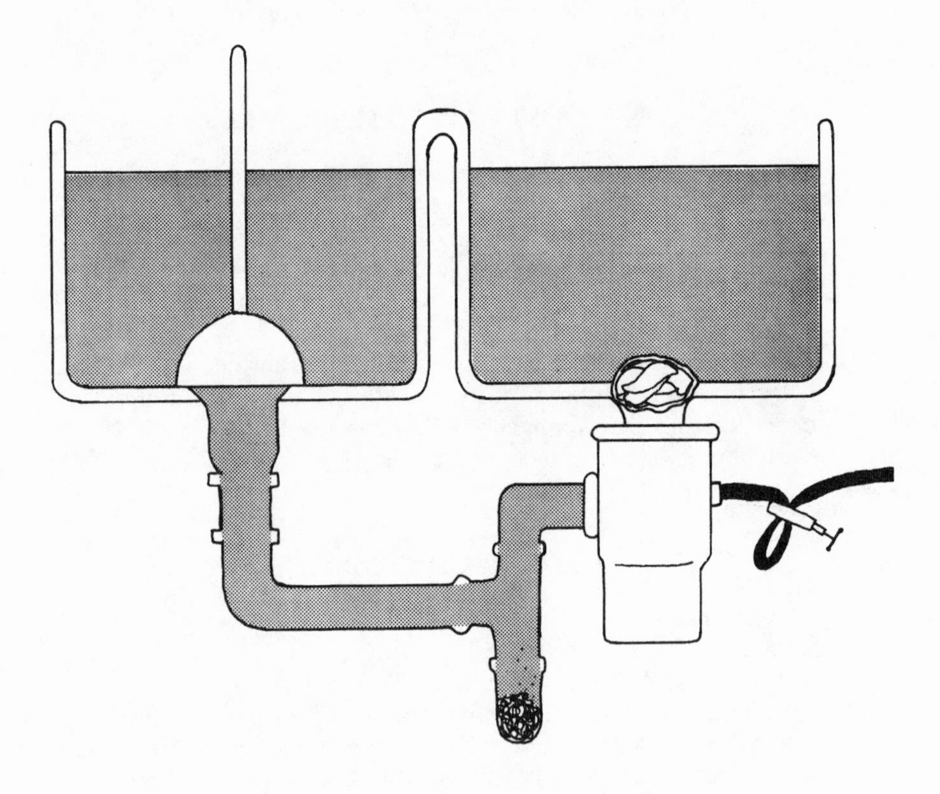

Double-compartment kitchen sink with plunger in place, the waste-disposer plugged, and the dishwasher hose clamped, showing a blockage on the vertical discharge.

in place. In many cases, a small woman's arms are not strong enough to hold the plug with one hand and plunge with the other, but a healthy, more athletic woman should have no real trouble pulling this procedure off.

If you have a dishwashing machine, you must plug the place where the washer drain hose comes into the sink drain. The easiest way to do that is to disconnect the piece of hose leading from the drain or garbage disposer, bend it over on itself and tie it off to hold it tightly closed. This will seal off the drain completely. Now plunge the sink. With all outlets stopped up, the plunger should do the job.

HELPFUL HINT: If you don't have enough slack in the drain hose to bend it in a "U" and pinch it off, another method is to use a wood clamp or a c-clamp with two small pieces of wood or metal to squeeze down on the hose and seal off the drain. With this method, you needn't bother to disconnect the hose.

The way the plunger works is to compress a column of water down on the blockage like a steel piston. It shocks the drain with many hundreds of pounds of pressure per square inch. In order to work at all, there can't be any other place for the water to go than right smack on top of the stoppage. Anything that lets water go anywhere else keeps the plunger from working.

When using the plunger, you must make sure that it is right over the drain and you must pump it very sharply, with much downward pressure, several times, to blow out the stoppage. Other methods will not work.

PLUNGING LAVATORY (BATHROOM SINK) DRAINS

In clearing a lavatory (bathroom sink) stoppage with a plunger, you use the same method as for the kitchen sink. The only difference is that you have to stop up the sink overflow drain near the top of the fixture. Rags and plastic are good, but where there are a series of holes in the ceramic, the best stopper is a piece of wood pushing down on a small kitchen sponge that is, in turn, pressing hard on a piece of plastic wrap. All the overflow holes must be covered or it's no dice. With your makeshift overflow stopper in place, and your

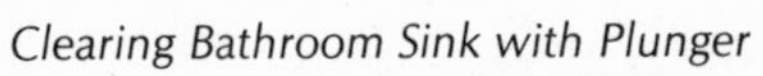
Clearing Bathroom Sink with Plunger

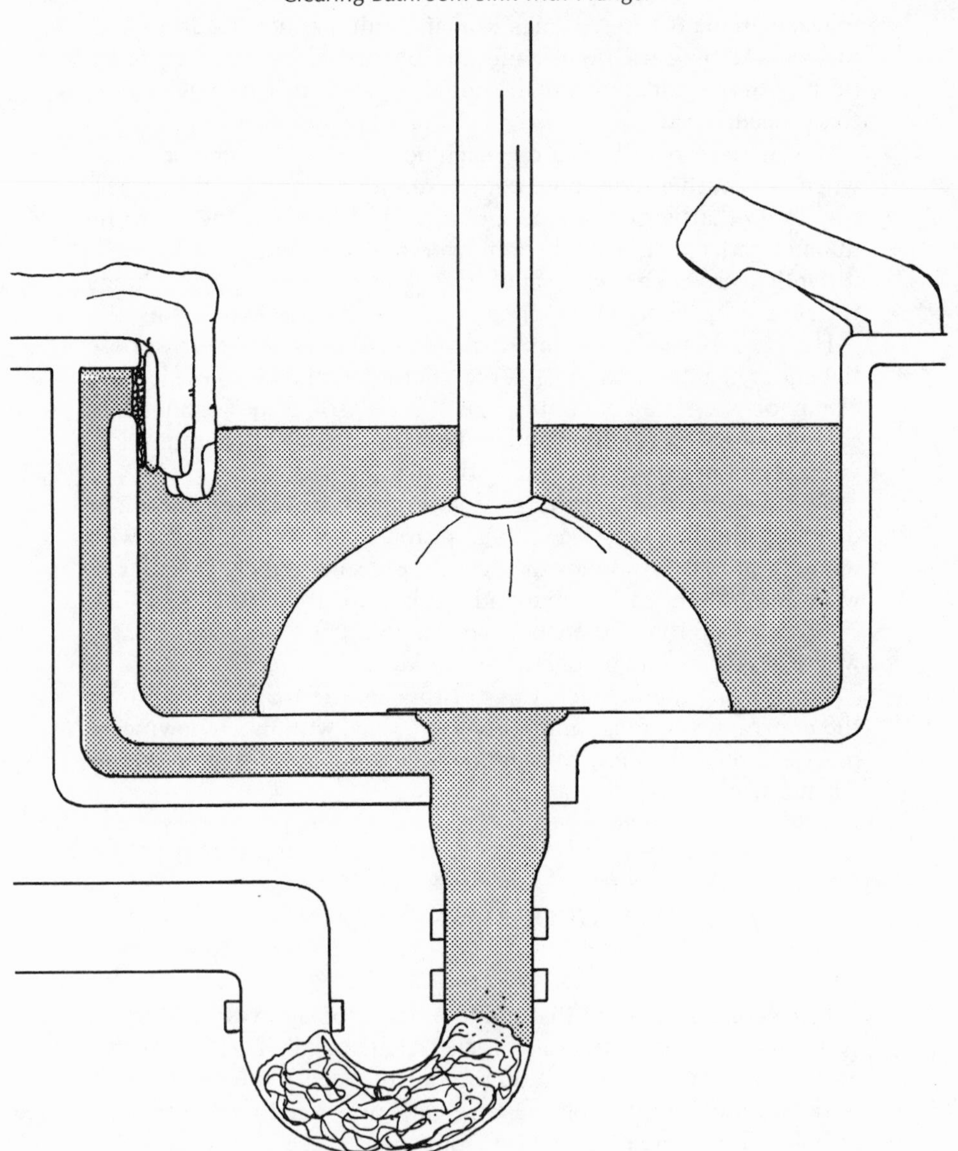

Plunger in lavatory (bathroom sink) with hand pressing improvised stopper over the overflow.

sink filled to just below the overflow drain, plunge vigorously. It should clear.

PLUNGING TUB AND SHOWER DRAINS

When plunging a tub, you must stop up the tub overflow as you do for a bathroom sink. To do this, you must screw off the overflow fitting and press your plastic-covered rag plug into the tube. Now, plunge away.

HELPFUL HINT: If your tub overflow contains the mechanism for the tub's pop-up stopper, simply pull the assembly out of the tube and set it aside. It is designed to be removed and replaced. It may be hard to take out and put back in place, but not to worry, pull and push as needed. It will come out and go right back when you're through clearing the stoppage.

Treat the shower the same way as you do the kitchen sink. Fill the shower base (pan) up to within a couple of inches of the top and plunge it vigorously. There are no other places for the water to go, so you'll be plunging the water column directly onto the stoppage in the trap.

THE PRINCIPLES OF MECHANICAL DRAIN CLEARANCE

If plunging fails, mechanical clearance is the only way to go. It takes longer, but it is not nearly as complicated and time-consuming as doing a toilet.

There are a few universal principles in this kind of work. First, with sinks of all types, you must take the sink's p-trap apart. That's the part usually made of chrome, but often plastic these days. Its shape suggests the letter "P." It connects to the pipe or pipes that run directly into your home's main sewer system, and is usually connected by means of large pipe nuts. These are called compression nuts, which means that they seal the pipe by compressing light rubber rings around it.

HELPFUL HINT: Before you even start to take the traps apart, go to your local supply house and buy some rubber slip-joint gaskets (rings). They can also be obtained in polyethylene (milky-white and translucent). If you've noticed that the chrome nuts are banged up or very corroded, it's a good idea to get replacement nuts. Kitchen sinks, laundry tubs, and some lavatories (bathroom sinks) take 1½" slip-joint nuts and gaskets. However, other lavatories require 1¼" nuts and gaskets. The best way to determine which to buy is to look at your kitchen sink p-trap and drain assembly. Then go into the bathroom and look at the trap assembly beneath the lavatory. If it is noticeably smaller, it is 1¼". *Caution: Be careful not to buy the cheap zinc die-cast nuts. They won't stand up.*

Some lavatory traps are 1½" x 1¼". One part of the trap (the part that connects to the main drainage system) is 1½" and the part that connects to the tube coming down from the basin is 1¼". Naturally, you can see the difference immediately.

To sum up: whenever you undertake to disassemble a p-trap or any of the chrome drainage tubing beneath any sink or basin, always have new rubber or plastic gaskets ready. Chances are about 90 percent that the old gaskets will either crumble or be so distorted as to be useless when you open up the system. If the p-trap itself looks very corroded or cracked on the outside, replace the entire assembly while you're at it.

The rule is to survey the problem and buy the replacements *before* you begin to turn the first nut with a wrench. These items are inexpensive compared to what you'd pay to have a plumber clear your blocked lines. A 1½" satin-chrome p-trap in 17-gauge shouldn't cost you more than $14 at the most. If you just need a few new chrome nuts and gaskets, the cost will be under $4.

After you've done any necessary shopping, get back to the job and wrench off the old slip-joint nuts. Slide them and their rubber gaskets up the pipe and out of the way, if possible. Then it should be a simple matter to drop the p-trap out. Be careful, the p-trap will probably be full of water, so handle it carefully. After you've emptied it of water, look at it. *Warning:* Don't empty it into the sink from which you've just removed it, for obvious reasons. Don't laugh; it's happened to me more than once, and it could happen to you.

In many cases of stopped-up sink drains, your inspection will reveal the exact cause of the problem.

If the p-trap is full of matted hair, grease, soap, pencil-stubs and a stray rubber lizard or two, the answer is perfectly obvious. Clean it out and return it to its place. Remember, when you replace the nuts,

Hold the P-Trap while Disassembling, to Avoid Spills

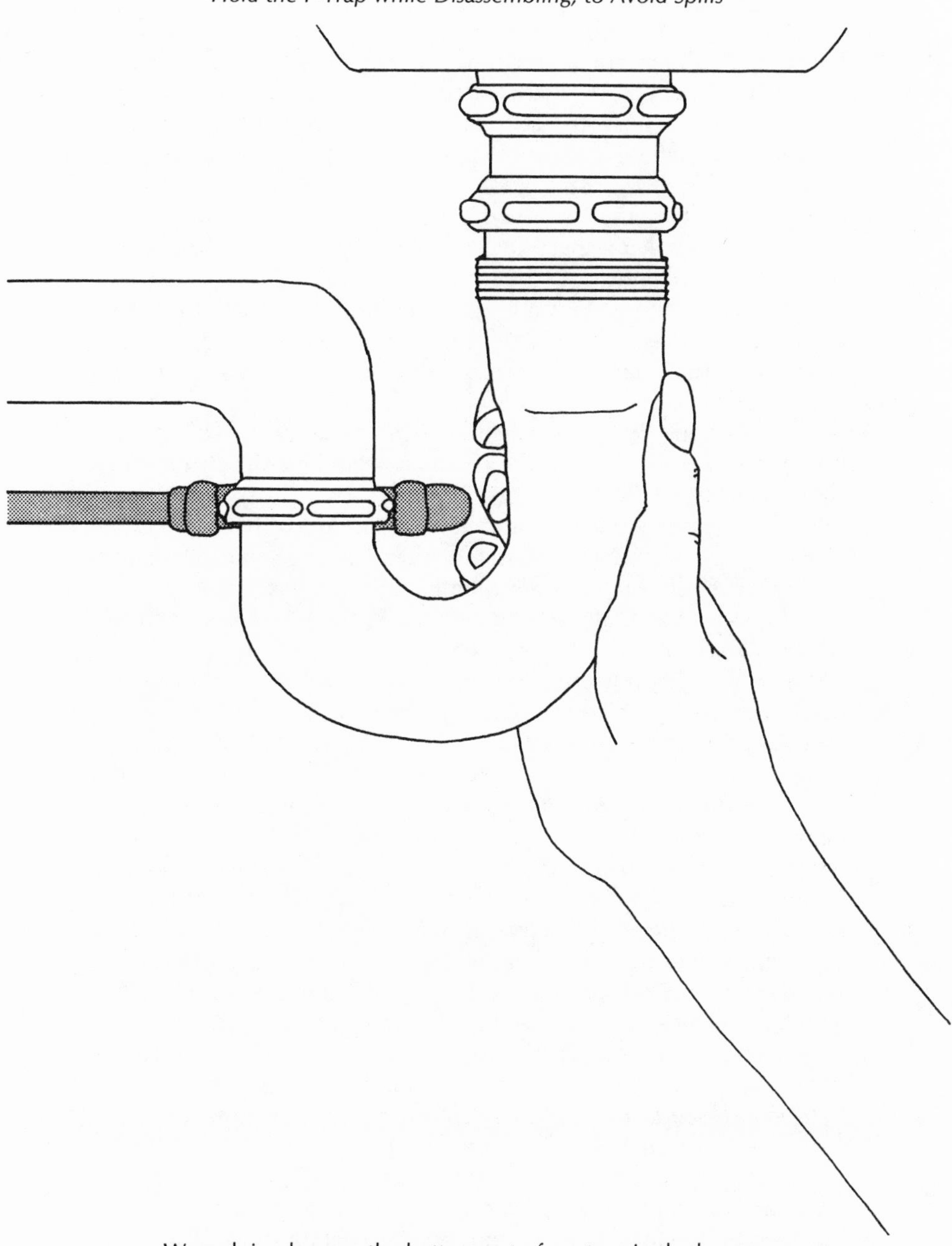

Wrench in place on the bottom nut of a p-trap in the loosening mode, with hand supporting trap to avoid spilling water upon disconnection.

not to tighten them too much. Make them *hand-tight,* plus a quarter turn with a wrench. *Do not, under any circumstances, tighten them more than that.* This is the general rule for all slip-joint compression nuts used in drainage systems. If they don't hold water, there are other remedies. If the joints leak when you run water in the sink for a test, simply take the trap apart again. Perhaps the square-cut gaskets have twisted, or there may be a defect in the parts (not an uncommon problem). A common cause of leaking is the misplacement of one of the gaskets on the p-trap itself. Normally, these rubber or plastic rings are placed on the tubing, after you've slid the nut up onto the tube. Then comes the gasket, so that when you slide the nut down to screw it onto the mating threads, it slides the gasket down with it and then compresses it as the nut is tightened. There is one place, however, where the gasket is seated on the fitting in its own special place.

Look at your p-trap, without the nuts and gaskets in place. The "U"-shaped portion has a short and a long leg. The short leg joins a piece of 90° elbowed chrome tubing which normally goes into the wall. Now look at that piece of tubing. At the elbow end, you will notice a rolled bead at the tip. The gasket is fitted into the space between that bead and the tip of the elbow section. There is just enough room for a rubber gasket. I do not recommend using a plastic one at this place.

With your gasket in place, you must bring the chrome nut from the straight end of the tubing down to the bead, which will act as a stop. The nut will cover the gasket. The reason for this arrangement is simple. That portion of the p-trap is a swivel, which allows for variations in the placement of the basin with respect to the drainage connection in the wall. The mating surfaces here are butted face-to-face, rather than the usual compression-nut-and-gasket arrangement, in which the nut squeezes the gasket against the side of the tubing, which slides into the p-trap socket. The long leg is adjustable for up-and-down variations and the short leg is adjustable for side-to-side variations. It's a very simple and practical system.

When you test your new or replaced p-trap, first run cold water for about fifteen seconds, full bore. Then, turn on the hot water and let 'er rip for a good fifteen to twenty seconds. This will cause the thin chrome-brass tubing to contract and expand rather quickly. If your joints are going to leak at any time, it will be at the end of this

Tightening the Top Compression Nut on a Typical P-Trap

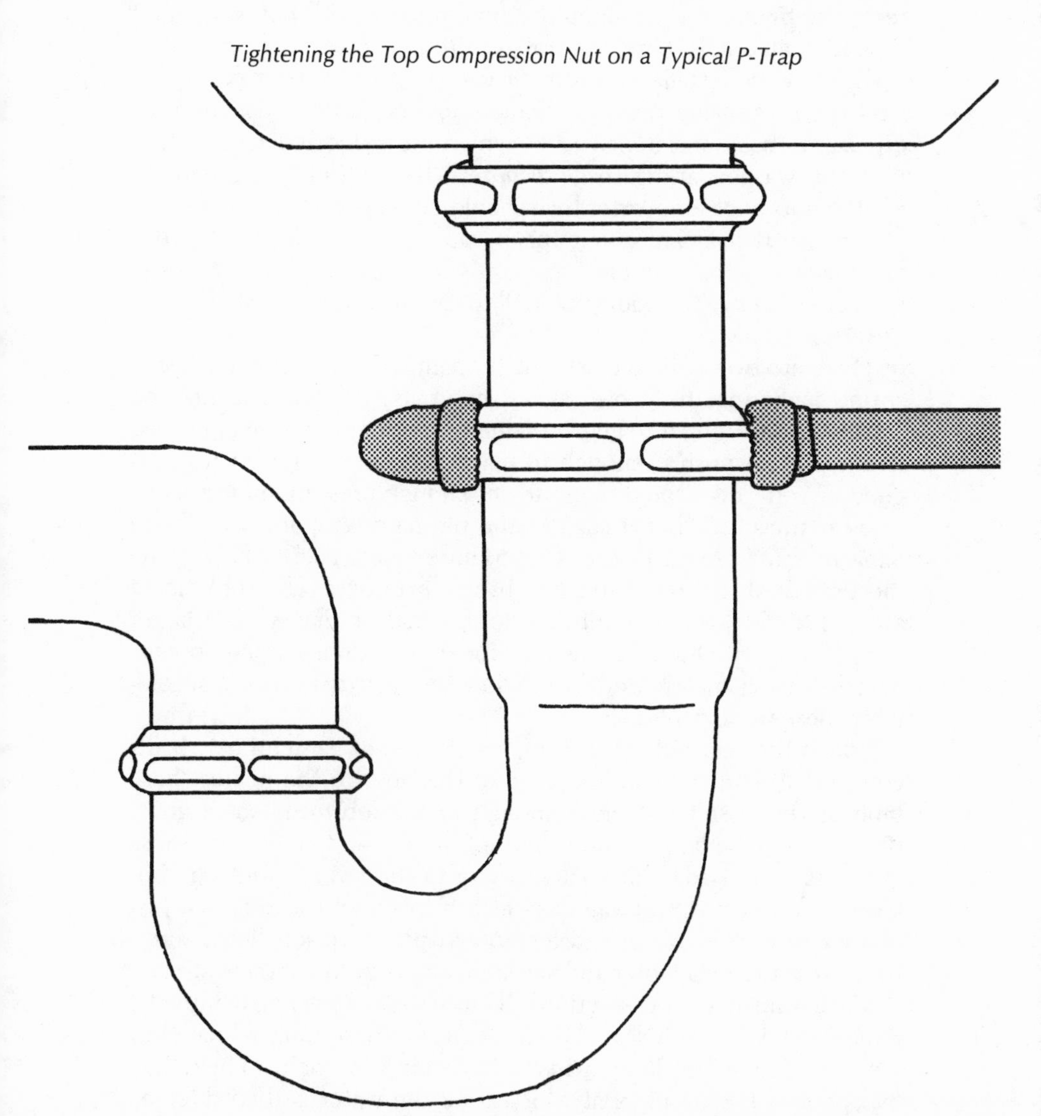

Wrench in place on the top nut of a p-trap in the tightening mode.

test. The final test is to fill up the sink or basin with hot water, and then let it all out. If there's no leak, you're home free.

Check your installation periodically, during the next few days, while you are using the sink. Sometimes the joints "shake loose," especially those on kitchen sinks which are subjected to vibration from dishwashers and garbage disposers. If you find a leak, tightening the nuts another quarter turn should do it.

Suppose that cleaning out the p-trap is *not* the answer to your blockage problem. The presumption is that you have a plug in the horizontal line going from your wall to the drainage-vent stack, or in the stack itself.

There are two ways to tackle the problem. The first is to take your spring snake (not the closet auger, naturally), and stick it into the stack, while turning it to give it enough movement to either grab the debris plug or push it enough to make it break up. Either result is good. If you find it too difficult to get enough pressure on the snake to clear the plug, then I suggest the ultimate weapon, your Drain King hydraulic drain opener (see Maintenance Tools). This is the most efficient tool for obstinate jobs I've ever used. Use the smaller size (Mod. 501) unit for tub, lavatory, and sink drains. The larger unit (Mod. 186) Drain King is best for shower drains and all sewers up to 3" in diameter. Carefully follow the instructions that accompany these wonderful tools.

Attach the selected Drain King to a length of garden hose. If it's convenient, connect the inlet side of the hose to the nearest hose bibb. If there isn't one close enough to the job to do some good, then it would be best to go to the supply store and buy one of those handy universal adapters that screw onto the swing spout on your kitchen faucet or to a lavatory faucet, in place of the aerators. Just screw the aerators off and fasten the adapter in place. Then screw the hose onto the adapter and you're ready to crank up the system.

Make sure that you insert the Drain King far enough into the drain to make a good seal. The principle of these tools is that they are designed to swell up when the water fills them. They then expand and fill the pipe, at which time the water is then directed through a spring-loaded nozzle in the tip. The water pressure causes the nozzle to pulsate, directing a hammer-like stream of water directly upon the blockage. The water pressure, plus this hydraulic pulsing action, will usually blow the stoppage into next week, clear-

Drain King 501 in Action Removing the Obstruction from a Bathtub Drain

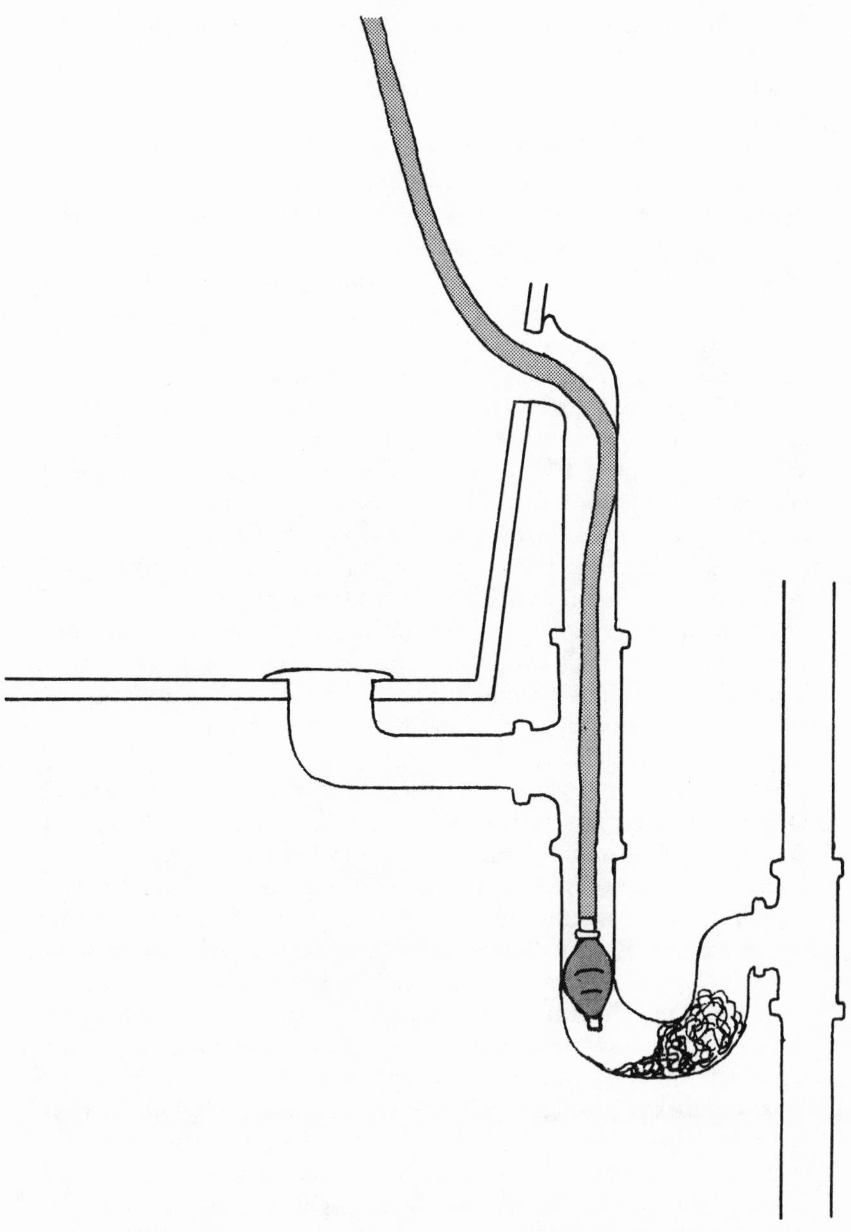

End-side section of bathtub, showing a Model 501 Drain King in place and fully expanded, just before the stopped-up trap.

ing the line. It is best to leave the unit in place for at least fifteen seconds after you hear it operating freely. Do it by the watch, because you don't realize how long fifteen seconds are until you've timed them.

In the case of the bathtub, the Drain King goes into the overflow tube, and is thrust as far down as you can possibly push it. The object is to push it past the tee-connection which comes from the tub drain, so that the stream of pulsing water is directed solely into the trap beneath the tub. If you see water gushing out of the tub drain, the Drain King is not far enough down into the overflow tube. You'll have to try again.

Important: Do not attempt to remove the Drain King and hose until you have disconnected the hose from the hose bibb or faucet. Just turning off the water won't get it. The bulb end of the tool will not reduce in size until the water pressure remaining in the hose is relieved by disconnecting it from the water source. *Do not try to tug on the hose with all your strength.* If it refuses to budge, give it a bit more time to deflate, then gently tug-and-release, tug-and-release, until you feel it coming loose. Then it's a simple matter to remove it.

Back to the tub. Now, reinsert the Drain King. As I said before, you must use the smallest unit available, or it won't make the turn in the overflow fitting. Remember, these things are high-quality rubber and can take a good bend, so don't be afraid of shoving away to get it in. You can't really do much damage to it.

HELPFUL HINTS: Coating the hydraulic unit with a bit of Vaseline often makes it go in a bit easier. Once you've made the turn, you can ram it home quite handily. It is important to use the thinnest profile plastic hose you can buy for these operations. Sometimes a bulky hose will resist the turn at the overflow fitting. When you've reached the bottom of the overflow tube, you should feel a solid contact. Now, try it again. If you still get water out the tub drain, there is one more solution to the problem. Get a rubber ball, about the size of a tennis ball or a bit smaller. If you can't find a smooth rubber ball, then by all means, use a tennis ball. Turn off the water to the unblocker, place the ball over the tub drain, put your foot on it and turn the water on once again. If you keep firm pressure on the ball with your foot, you should be able to maintain a strong internal pressure in the pipes, which may be enough to clear the blockage. The principle here, as it is with the plunger, is not to allow the water to go anywhere except right to the blockage. If it can escape to any other place, of course it will, and you'll be wet, tired, and discouraged, with thoughts of calling a combination plumber-banker in to administer the *coup-de-grâce* to a problem that you've nearly licked. All that would have been needed was a bit more patience and understanding of the principles involved. *Remember:* If you can bring overwhelming pressure to bear on the stoppage, it will clear. Every plumber knows that. You know that iron or plastic pipe isn't going to give. You should realize that the water in your house supply system is at least thirty-five pounds per square inch pressure, and usually as high as forty to sixty pounds, or more. It stands to reason that no blockage, unless it's plaster or cement, is going to resist such pressure.

Which brings me to a very special place in this story. *The most devastating types of blockages in drains and sewers are the result of pouring solutions containing cement, plaster, patching compounds, latex paint, water-soluble glues, casein, and other hard-setting compounds into the house plumbing. If you must do it, make sure that the material you discharge into the drains is diluted at least ten to one, that* cold *water runs full force at the time of disposal, and that you allow the water to continue to run for at least three to five minutes after the material has gone down the drain, to insure that it reaches the street mains, and isn't allowed to build up in your house plumbing. Do not discharge these hard-setting products into your house drains on a regular basis.*

I have seen house sewers that have had to be removed entirely and replaced at a cost of thousands of dollars as a result of do-it-yourself improvement projects, which have saved householders a few hundred dollars. *Dispose of hard-setting materials into pits in the earth, or pay the few dollars required of disposal in public dumps. In many cases, your regular garbage service will be adequate to the task.*

I cannot overemphasize the importance of using your head in such matters. In plumbing, as in anything else of enduring consequence, common sense and knowledge are the keys to success.

Again, in regard to unblocking with the Drain King, use the smallest model on your bathtub drains, and either it or the next size up on your lavatory and sink drains. The larger unit is suitable for clotheswasher standpipes, which are the simplest drains to clear, since all the plumbing is open. The trap is usually attached directly to the standpipe, and its vent is immediately behind or adjacent. Washer standpipes are not able to be plunged, so the only alternatives are snaking or the Drain King. Frankly, I prefer the Drain King.

And then the day comes when you've done everything you can, everything I've told you to do in this book, and you're still blocked up tighter than a tick.

You think all is lost. You're in for a battering at the hands of Fagin-le-plumber. Not so. First, the most important thing to establish is the nature of the stoppage. Is it concrete, sort-and-curlies mixed with bath soap, Sis's long blonde hair, your rug rat's plastic teething ring, or Dad's weekend do-it-yourself project, complete with greaseballs, mechanic's soap and mop strings? Well, frankly, if you've done all the things we've discussed in this book, the chances are that what's in the drain is either very durable and persistent like plastic or metal or it's hard-setting material. The Drain King would be ineffective, so it's time to go back with the snake and start wearing the object away, by persistently turning it against the blocking material. If you find that you have difficulty getting your hand-driven snake down the drain, then, by all means go to your local do-it-yourself or contractors' equipment rental store. Every major metropolitan area has these stores. Tell the clerk the size drain you are trying to clear so that he or she can prescribe the proper tool or auger size. Rent a motor-driven snake, and do it like a pro, which, by this time, you should be, compared to your less-enduring neighbors.

If a motor-driven auger doesn't do the trick, you're probably in for it. Your last recourse is the professional drain-clearance specialist. They usually have the word "Rooter" in their trade names.

HELPFUL HINT: If it's the drain beneath the toilet which needs clearing, have the toilet off and the open flange available to the operator's inspection and use. These outfits usually charge extra for removing the toilet and resetting it. On your other drains, to save yourself some possible damage to your chrome-plated p-traps and fittings, take them apart yourself. If you save the "Rooter" person time, you will save yourself some money.

UNBLOCKING THE MAIN HOUSE SEWER LINE

Of course, if you find that your entire house drainage system is

Spring Snake Snagging a Drain-Line Stoppage

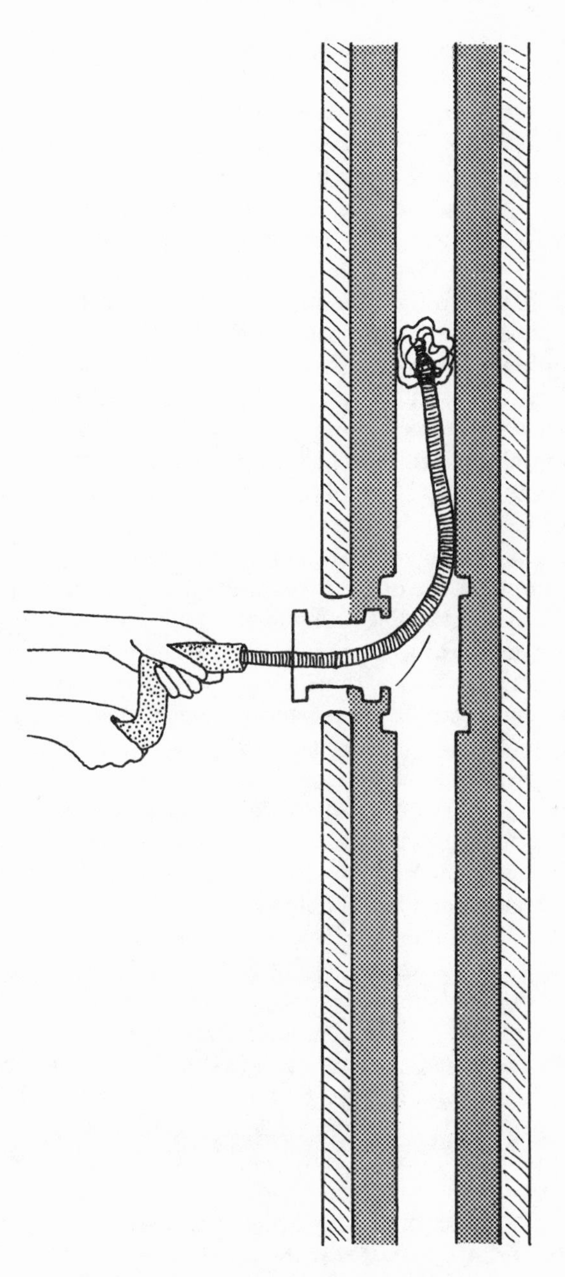

A section of pipe in the wall in which a spring snake has been inserted and has buried its tip in a blockage, in preparation for removal.

stopped up, you realize that it's the main line going to the street which needs the attention. In cases of very old homes, the mains are made of materials that can crack under the stress of the earth movement, which can be caused by any number of phenomena, including a lowering or rising water table, earthquakes, land-fill shift, flood, drought, or the repeated movement of heavy construction equipment nearby. Cast iron, clay, asbestos cement, bituminous and cast-cement pipe can and usually do crack under such circumstances. The fluids going through these lines contain a form of natural fertilizer, to say it delicately, and this seeps into the soil, attracting roots, which eventually grow right into the pipe, seeking a continuing supply of rich nutrients. That is why I usually suggest the introduction of organic weed-and-root-removal treatments into the drainage lines of older houses, on a regular, programmed basis.

Believe it or not, the best place to get information on such chemicals is your local janitorial supply house, especially if the house sells sanitary chemicals, solvents, and other maintenance products. If the janitorial supply house doesn't have the kind of product which I recommend—a biodegradable root-and-weed killer such as those recommended for unblocking drains in homes which are served by individual septic tanks systems—then, by all means, call up your nearest general chemicals supply house and ask one of their marketing people to recommend a product and tell you where to buy it locally. They'll usually be very cooperative.

The stopped-up main usually means professional "Rooter" clearance, since the equipment used is designed to cut out the roots to the inner walls of the pipe, and it is unlikely that you will be able to rent such an auger at a reasonable price.

Before you resort to the "Rooter" however, there is one option left to you. If you want to take the chance, go to the equipment rental store and get the largest motor-auger they have. Make sure it has a spring rod long enough to reach from the farthest back cleanout in your house sewer main to the curb or other location of the city main. If you're on septic tank, you should know the distance from the cleanout to the input side of the tank. Even though the rented motor-auger may not have the bit diameter to clean out your pipe to the walls, it will probably have enough sock to punch through the obstruction in the main. When that happens, the best thing to do is run cold water with added compounds (organic drain unblockers and weed-and-root controls) through the lines from every fixture in your house.

Do this, on a planned basis, once each week, religiously, for the next month or so, and you should bring the problem under reasonable control.

If you notice the drain flow getting significantly less during the next month, even with treatments, it usually means that you have a really persistent weed-root problem, or that concrete or other hard-setting compounds have built up to critical levels in your drains. In this case, you have no option but to get the "Rooter" man on the job.

Extreme caution must be exercised in the use of unblocking compounds and root-and-weed killers. They are active poisons and should be handled with gloves, and all fixtures into which they are introduced—especially sinks, lavatories, and bathtubs—must be rinsed and cleaned thoroughly after the application of such products through their drains.

You may say that, in the end, having to call the professional has cost you more money and grief than if you'd called him in the first place. In a way, you're right, but, in another, more important sense, you've paid for an education which could stand you in excellent stead later on.

Now you know the nature and severity of your problems. You are starting from square one again, with cleanly scraped pipes. Using preventive maintenance procedures described herein, you're on the way to a long-range cure of your problems, and, in the end, where others have not profited from their experiences, you have. They may spend between $50 and $80 per year, *every year*, for continuing "Rooter" remedies, but you'll have a one-time expense and the prescription for maintaining your system at minimum annual expense.

In the final analysis, your knowledge of your home's plumbing system should give you sufficient confidence to consider even more elaborate alternatives—including replacing badly rooted or deteriorated plumbing—later on. If that comes to pass, this book will talk you through that process as carefully and completely as it has done with preventing and clearing blockages.

LEAKS

There are two basic types of leaks: pressure leaks from your water or gas supply lines, and leaks from your drainage system.

The most common are pressure leaks. They most frequently occur in faucets, valves, and galvanized-steel water lines.

Galvanized steel pipe is designed for a service life of seventeen years. This means that, under normal conditions in your home, American-made galvanized steel pipe should maintain sufficient inner dimensions, wall strength, and resistance to normal wear-and-tear in continuous, daily water-transmission service to deliver its design capacity for that length of time.

I have removed galvanized pipe that has been in service for over fifty years and still has a reasonable waterway and strength. I have also torn out galvanized pipe from fifteen-year-old homes that is so deteriorated that you couldn't pass the lead from a pencil through the corrosion buildup in a ¾'' pipe.

Causes for rapid deterioration are: (1) "electrolysis" (properly called galvanic corrosion), (2) water having high mineral content or transmitting abnormal amounts of undissolved solids, (3) highly acidic or alkaline water, (4) poor plumbing installation, with excessive amounts of 90° fittings in sequences which make them act like traps, (5) pipes installed in ground that is highly basic or acidic and that attacks the outside of the pipe.

Some of these causes and effects will be discussed in the sections on installation of new water and drainage pipes. We will only touch on leak remedies in this section, which should cover your immediate concerns.

Leaky faucets and valves are the biggest pain in the butt of all, because you see them every time you look at the offending fixture. They drip and cause disturbances in the night. They are the first plumbing problems most people have learned to remedy. There are some you can fix in a jiffy, and some that will require walls to come down and major pipe operations to be inaugurated to eliminate the causes.

Fundamentally, there are three classes of faucets: ball-type, washer-type, and washerless. There are basically three types of valves in common residential service: the gate valve, the globe valve (to which most classes of faucets belong) and the ball valve.

When you turn a gate valve handle clockwise, it lowers a metal disc-type gate of the same composition as the valve itself into the waterway, shutting off the flow. As you look through an open gate valve, you can see straight through it. On the other hand, a globe

Gate Valve

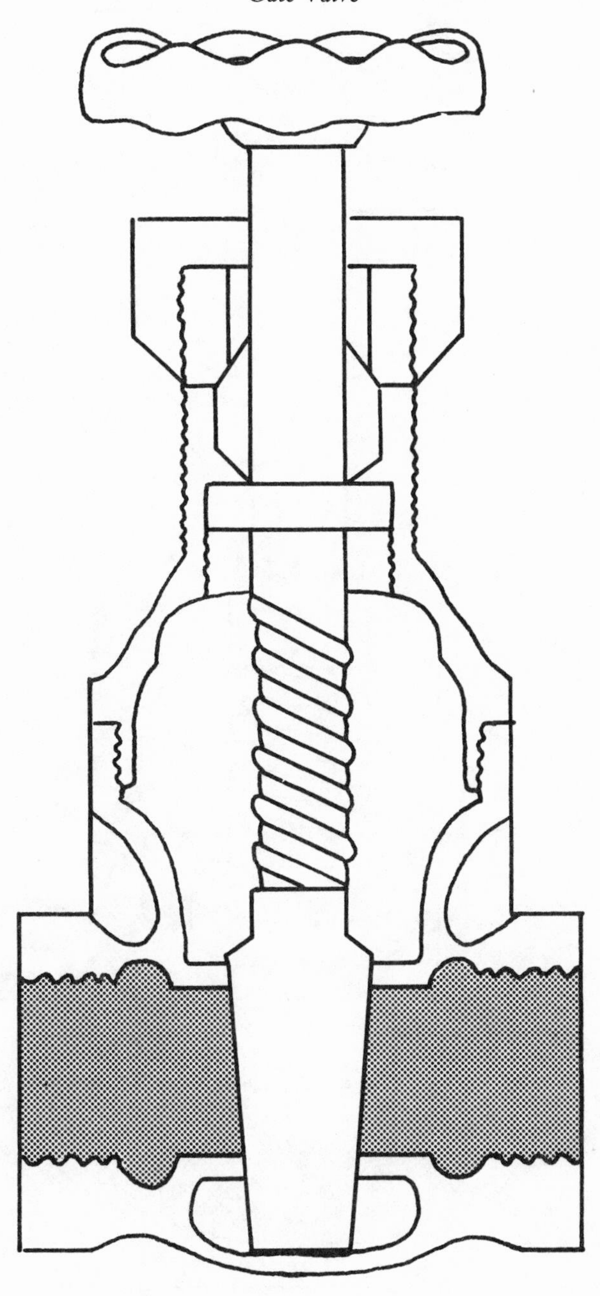

Cross-section of a gate valve. The closed gate looks like a keystone.

Globe Valve

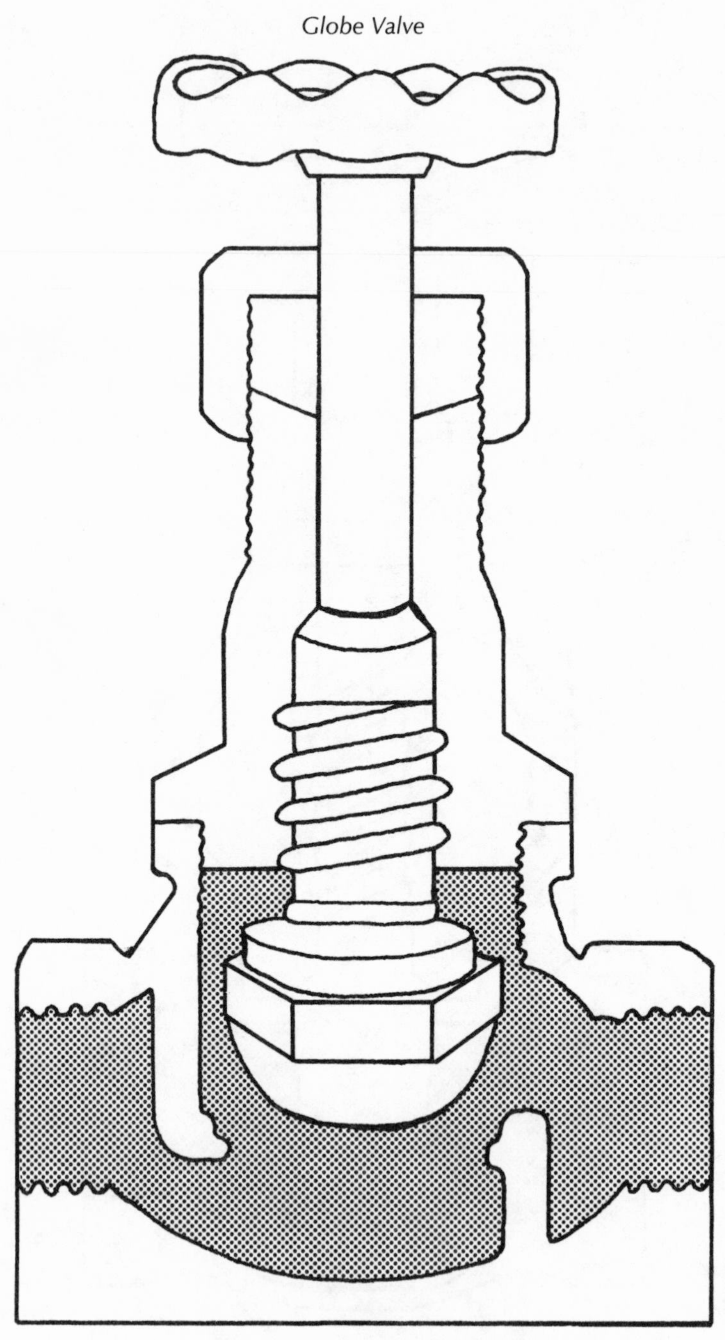

Cross-section of a globe valve. The partially open stem looks like an upside-down derby hat.

valve has a composition disc mounted perpendicular to the stem which closes down on a circular seat to stop the flow. When you look through this valve, you see only the valve casting. The angle valve is a globe-type valve. The difference is that the water comes into the valve from the bottom and goes out from the side (a right-angle flow pattern). Gate, ball, and globe valves are straight through. So, now you see that the small chrome-plated valves beneath your wash basin are angle valves or "stops," valves of the globe family, as are your faucets. Hose bibbs can be either angle valves or straight-through globe valves. The ball valve is simply a valve body with a handle-operated ported-ball seated in it. The handle turns the ports or holes to inlet and outlet channels.

Gate valves very seldom need replacement gates or metal discs. They are designed to open wide or close completely. They are not designed to throttle or reduce the flow of water or gas. Globe and ball valves, by their design, can be used to control the exact flow of water.

The globe valve's composition disc closes against a ground seat which gives infinite variation. But, because of this very characteristic, the discs in this type wear and need periodic replacement when they leak.

Replacement of faucet washers is simple. My list of maintenance tools includes an assortment of faucet washers. Pennies will save you many plumbers' dollars.

FAUCET LEAKS

If it's the kitchen faucet that's leaking, and if it is of the standard, two-handle variety, mounted on a sink deck, the procedure is fast and easy. Turn off the chrome-plated angle stops beneath the sink. Turn the handles on the faucet to take the pressure off, and just look at the faucet carefully before you do anything more. Obviously, you must remove the handles to get at the washers. Some sink centersets have very simple handle designs. You can see the screws right there in the center of each handle. Remove them and give the handles sharp tugs. If they don't come up easily, then get your plumbing-fixture faucet-handle puller, open it up, set the pin in the screwhole in the middle of the handle and the two jack hooks or blades beneath the handle, and turn the puller screw, thus lifting the

Typical Deck-Mounted Mixer Faucet, Showing Parts

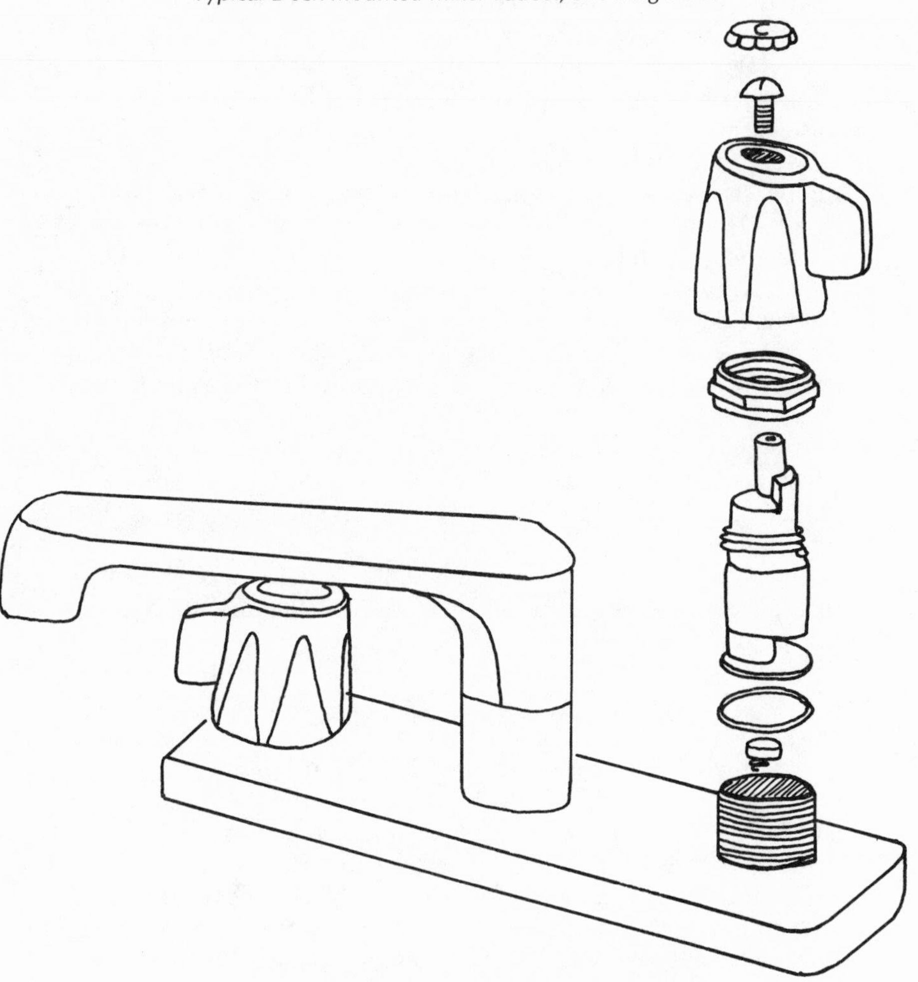

Lavatory (bathroom sink) faucet with cold-water (right) valve disassembled.

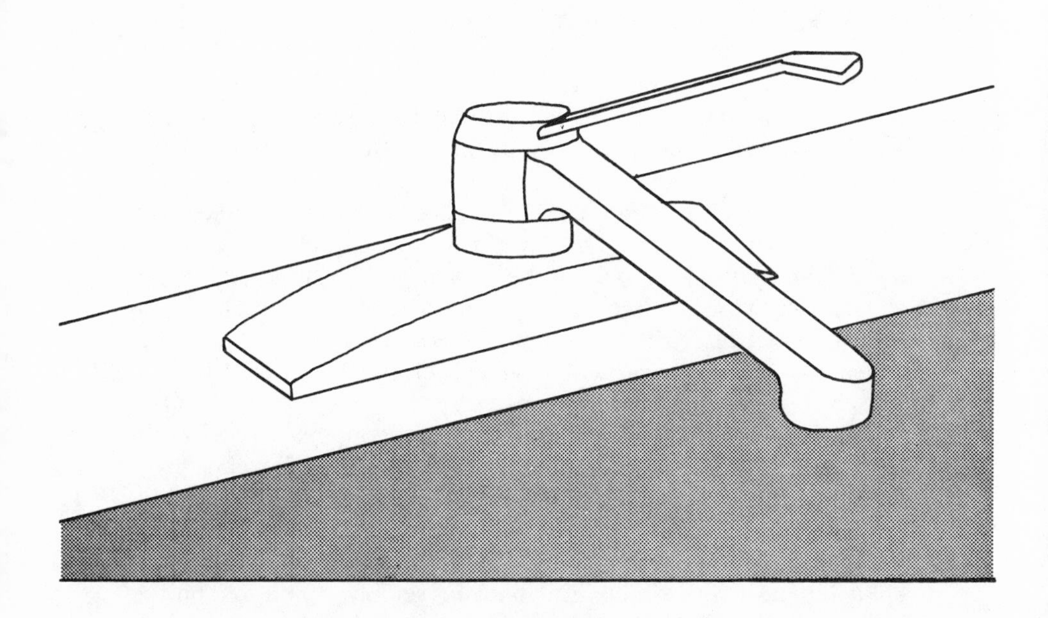

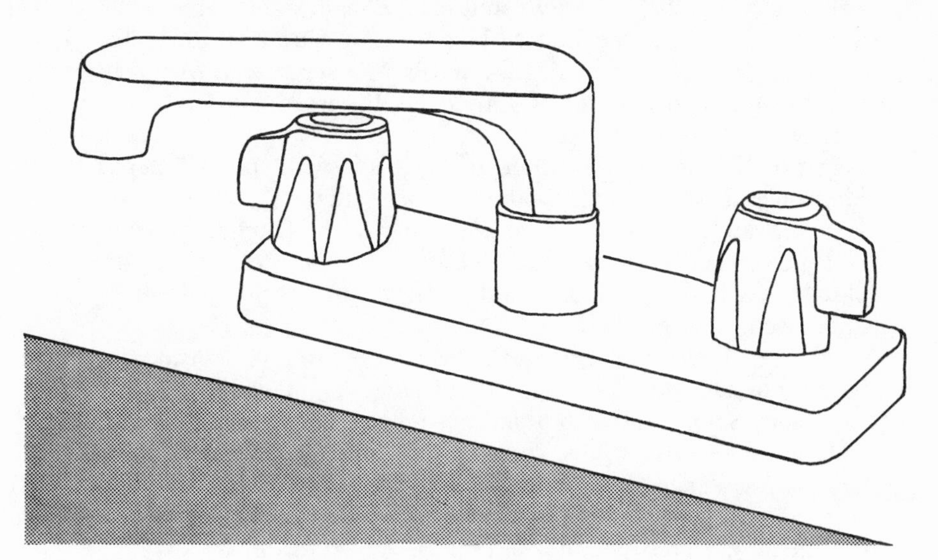

Top: single-handle sink faucet; *bottom:* double-handle sink faucet.

handle gently from its shaft. Repeat the procedure with the other side.

Removal of the handles will probably disclose the stem-packing nuts and the bonnets of the valve assemblies. It there are two hex nuts (hexagonal-shaped nuts), simply open the larger of the two, the nut closest to the valve body. The nut closest to you tightens the valve packing around the valve stem to keep the water from leaking while the valve is open. When you've wrenched out the stem and disc-holder assembly, you'll see the washer on the inside end of the stem. Examine it carefully for wear and deformation. While you're at it, take notice of the condition of the screw holding it in place. Often the screw's driver slot is badly deformed or worn. If it is, replace it when you replace the washer.

The very best types of washers to use are neoprene-beveled washers, rather than the flat red or black rubber type. The beveling gives greater seat contact and the neoprene gives a longer service life. However, if all you can find for your maintenance kit is the flat rubber type, so be it. Use what you can get.

Important: If your kitchen sink faucet assembly is mounted on the wall, you will probably have to shut off the main water valve, wherever it comes into the house. This type of faucet is connected directly to the pipes in the walls, and has no separate valves of its own to shut off during repair. Otherwise, the wall-mounted type is exactly the same as the deck-mounted centerset.

At this time, I must mention the type of swing-spout-centerset sink faucet that has handles with concealed screws.

If you can't see the screws in the center of the handle, they may be concealed beneath the *H* or *C* discs set into a recess in the handle. It is usually obvious whether they can be popped off or unscrewed.

On the plastic crystal handles, the concealing round decorative button normally may be popped off by gentle prying with a blade of some sort. Some American Standard and many of the older Kohler and Crane centersets have screw-off buttons. Generally, this type has a knurled edge like a dime, and will turn easily with a small pair of pliers or other light jawed tool. If they aren't too soap encrusted, they might even respond to your fingers.

HELPFUL HINT: Faucet washers come in a variety of sizes, so at first get a selection, until you learn (through the experience of serv-

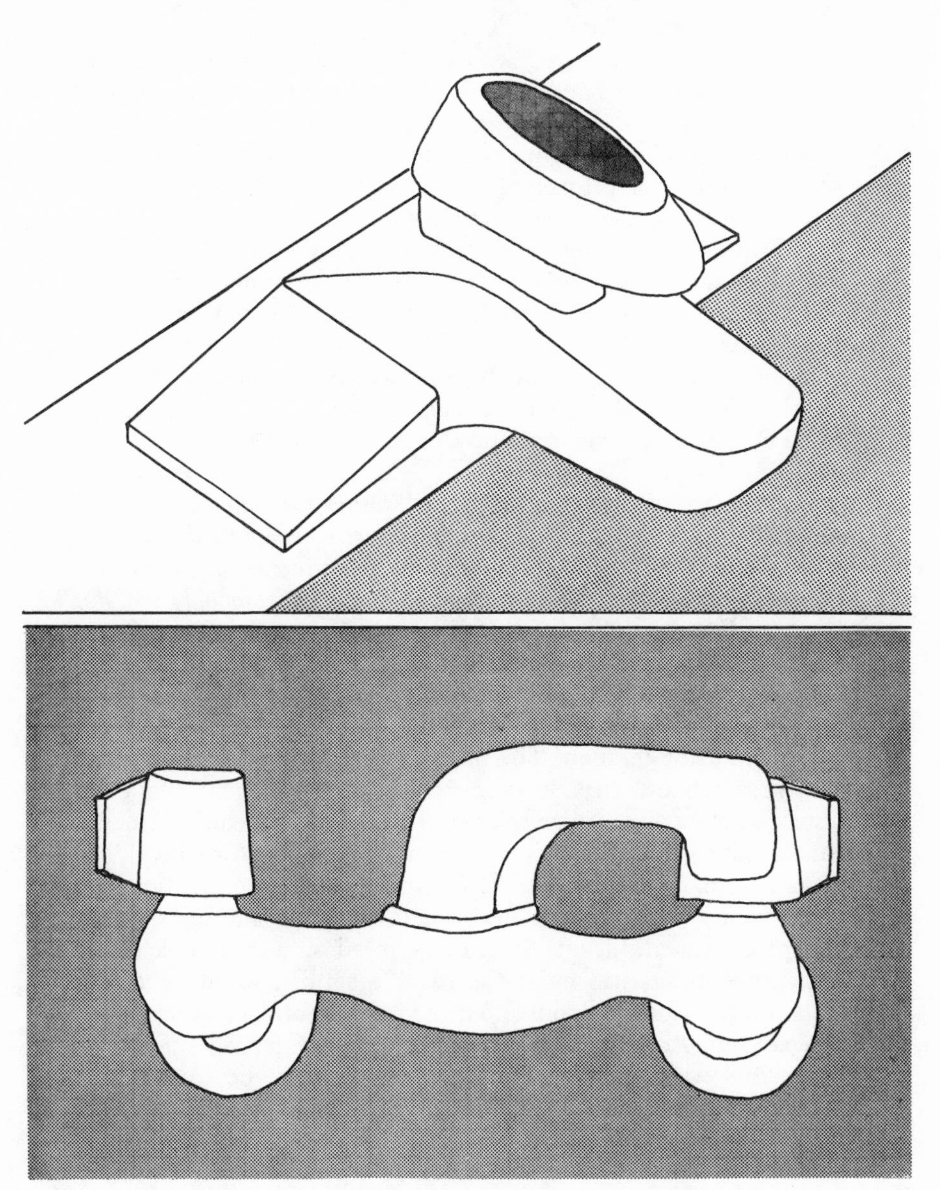

Top: single-handle lavatory (bathroom sink) faucet; *bottom:* wall-hung kitchen faucet.

icing your fixtures) the sizes which each faucet requires. Then, just keep those sizes on hand. Faucet washers are so inexpensive that keeping the surplus should pose no financial burden, and, one day, you may replace an existing faucet or bibb which could require washers from your surplus stock. You can always give your excess washers to needy neighbors as a goodwill gesture, but chances are that if you live in a so-called tract house, with many similar homes in the vicinity, every one of them will have been plumbed identically, and your neighbors will need the same stock sizes as those that fit your faucets.

In most cases, unless your fixtures have been modified, all bathroom faucets—lavatory, shower, and tub—will have taps manufactured by the same company, and should have identical replacement washers.

Warning: When selecting the proper washer for your faucet, make sure that you don't force it. The stem-washer cups are made to accept precisely the correct washer. You must not try to force a larger size onto the cup, and a sloppy-fitting too-small washer will probably not stop the water flow completely.

Before replacing your newly washered stem back into its proper place, get your flashlight and look into the body of the faucet. Take note of the raised ring onto which the washer presses to make the closure and stop the flow. The ring will usually be shiny from use. If it looks rough and dirty, or if your old washer appears shredded or otherwise torn up, you may have a bad seat, in which case you should not replace the stem until you've made the seat smooth and even once more. The faucet-valve grinder which I've mentioned in your maintenance tool kit is designed to smooth rough seats. It consists of a serrated, file-quality disc, mounted on a tee-handle. Obviously, the object is to move the rough steel disc against the seat, until you have evened it out and made it as smooth as possible.

Note: Some of the more expensive types of faucets and valves have seats which can be removed. The clue is the square hole. When you look down at the seat, if your seat-ring surrounds a square hole instead of a round one, it may be designed to be removed with a square wrench or driver stud. Often, a screwdriver will do nicely, but if it doesn't budge after a careful try or two, just go ahead and grind it as if it were a fixed seat.

If it's removable, chances are that the local outlet for that brand of

Grinding a Valve Seat

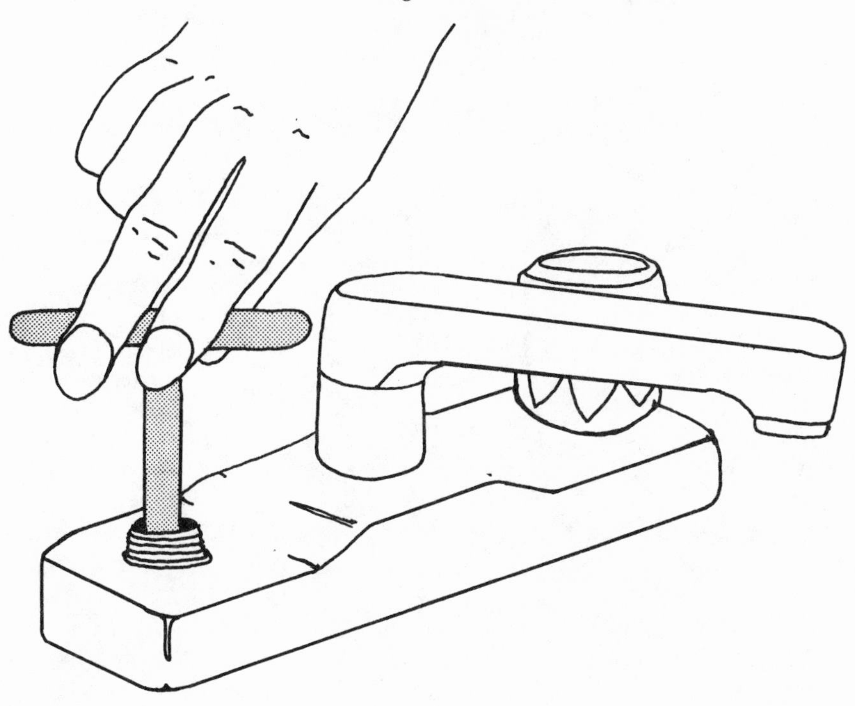

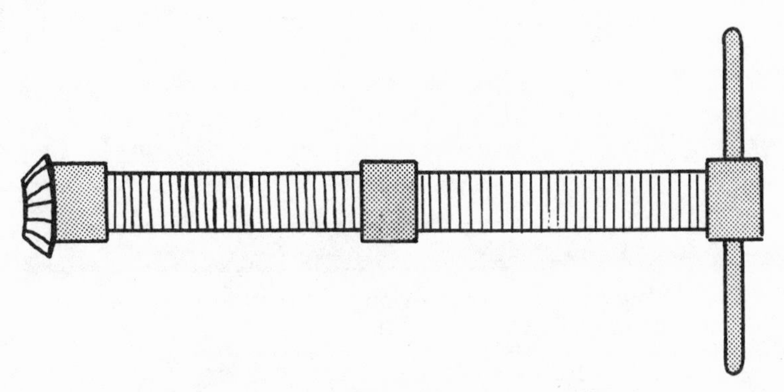

Hand turning a tee-handle faucet-valve grinder in a lavatory (bathroom sink) faucet.

Turning Out a Removable Faucet Valve Seat

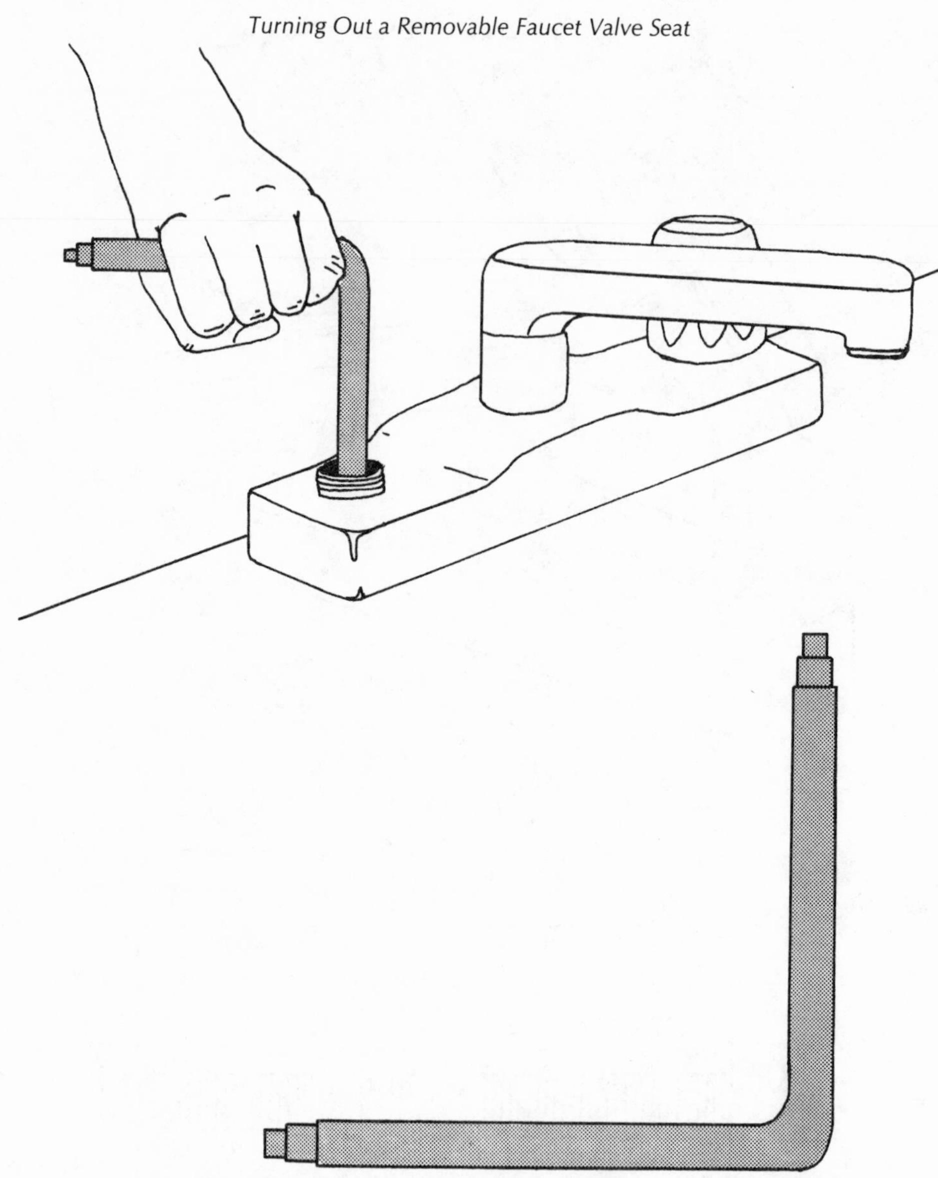

Hand turning out a removable valve seat from a lavatory (bathroom sink) faucet.

faucet will have replacement seats, and you'll be starting from square one, with a new washer and new seat, which should give the faucet exactly double its original life.

On the fixed-seat type, seat replacement is impossible. If you see black lines across the ring of the seat, gently reach in with your small finger and try to scrape your fingernail around the seat. If your nail drops into the line or lines, the seat is wire-drawn and will leak water no matter what you do. If grinding doesn't eliminate the lines and depressions, you will have to replace the faucet. No amount of stem pressure will make the faucet tight enough to eliminate leaks.

Obviously, if the faucet seat is broken or seriously cracked, and it can't be replaced, the faucet is "kaput."

The principles which I've described for sink-faucet repair would naturally apply to your washer-type lavatory faucets as well.

SHOWER VALVE LEAKS

Your shower valves are similar but there are a few special wrinkles I'd like to discuss.

After you have removed the handles, and you've unscrewed the escutcheons (those chrome or ceramic dome-like decorative parts against the wall), you may discover that the larger of the two brass hex nuts is below the surface of the tile. *Ouch!* You say to yourself, how in the name of that great master-plumber in the sky am I going to get the stem out? Good question.

The only answer is that you are going to have to do whatever it takes. That usually means that you'll have to chip out some of the tile around the stem. Old tile is as brittle as glass, and any undue pounding will usually ruin it. The damned thing will possibly crack right across its face, the moment you begin your chipping.

The answer to this problem is to try the best you can to locate matching tile *before* you begin your project. If the tile is too old or faded to match exactly, then try to find some contrasting tile, and resign yourself to removing more tiles from the shower wall, if necessary, to form a neat square pattern behind each valve. There are other interesting variations, as well, including a band of contrasting tile across the wall.

But let's hope that new tile won't be needed; that you're so careful and lucky (emphasis on the latter) that your magic hands chip

Shower Valve with Parts Exposed

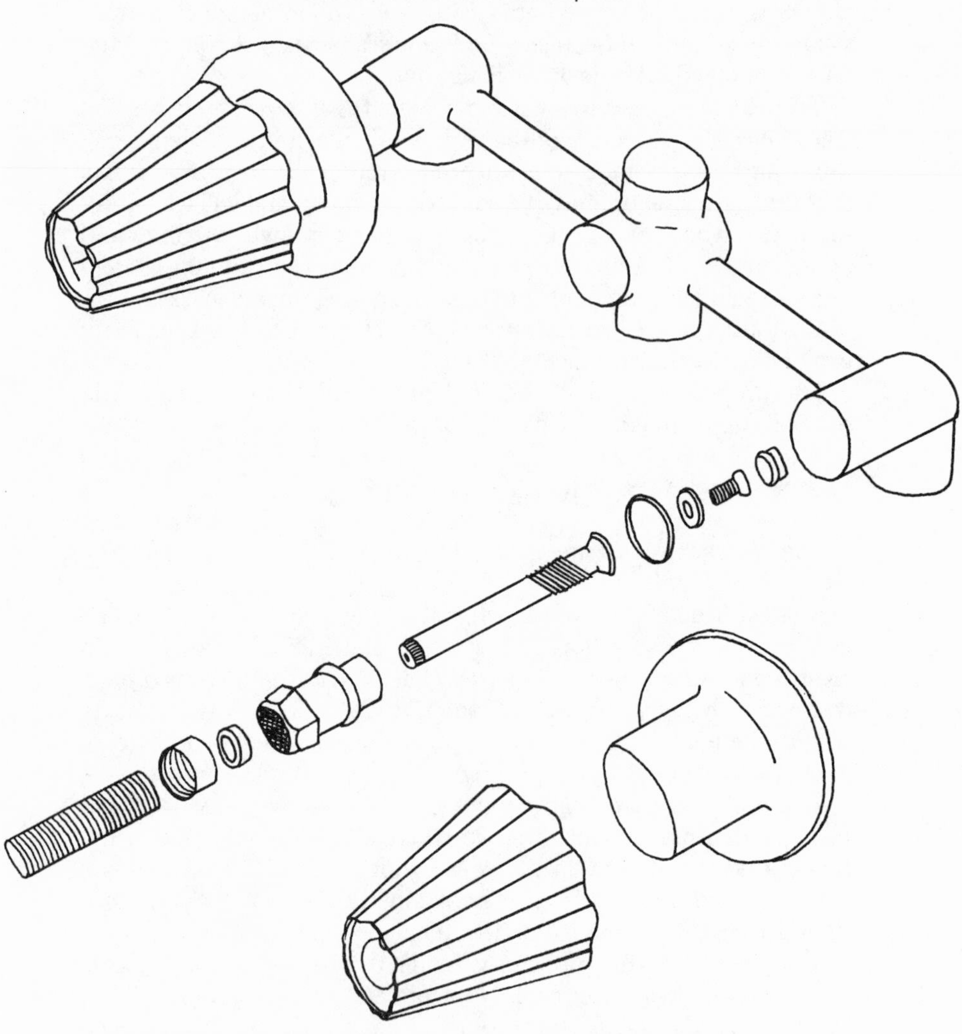

Shower valve with parts exposed. The knobs are fluted and tapered. It is obviously intended to be mounted on the wall. The distinguishing feature is the tee-like assembly between two rather thin tubes, which are attached to the two valve bodies. There are top-hat-like chrome wall escutcheons.

out just the correct amount of tile around the valve stem and the bonnet hex nut to give you a grip with a deep socket of the proper size.

By the way, if you have a ½″ ratchet drive-and-socket set, which I haven't mentioned in my tool selection, the addition of a deep socket of the proper size to handle your shower valves would be a valuable addition to your equipment list. You may find yourself using it more often than you think, since it will also probably work on all other faucets of the same manufacture. In any case the deep socket may save further tile chipping that can occur with larger wrenches.

Having chipped out the tile sufficiently to permit the use of the socket, but not to such an extent that your work will show when the escutcheon is returned to its proper place, simply turn the large hex nut out of the shower-valve body. The rest of the work is identical to the procedures described for washer-type faucets.

Suppose you find a bunch of plaster or concrete around the stems of the shower valves? The answer is the same as before. You must get to the bonnet nut to do any good at all. That means chip away, no matter what the consequences. You can always repair the wall, but you can't fix your shower mixing valve unless you can get into its guts. Overcome your insecurity and/or fear and get on with it. The time to consider repairs to the wall is later.

HELPFUL HINT: If you haven't a ½-inch drive, ratchet set, or even a ⅜-inch drive, perhaps this would be a good time to invest in one. If that doesn't appeal to you, then try your 12″ channel lock pump pliers. The curved jaws will probably reach in and grasp the nut. The problem will be putting enough torque on the nut from this awkward position to get it free and turning. Another alternative is to chip out enough tile to get in with an end wrench or crescent.

When you get the stem out, be careful not to let tile chips, concrete dust, and grout particles get into the valve body. They could get on the valve seat and keep it from closing. Such abrasive particles in the body also could wear out the stem threads ahead of their time.

Your first big shock may come if and when you discover that your shower mixing valve seats are shot to hell, and are not the replaceable type. The answer: Put it back together with new washers temporarily. Then plan one whole weekend around installing your new shower mixing valve. It's a major project.

Washerless faucets are a boon to mankind! They seldom blow, but when they do, most manufacturers will have provided repair kits with easy-to-follow instructions. Another advantage attached to their use is the fact that there are so many inexpensive washerless mixer faucets on the market. In fact, they are so cheap that, in many instances, the simplest solution to your problem is the total replacement of the offending unit.

Remember that a new faucet will probably cost less than a plumber's wages for one hour. I have seen very few of these faucets go down in less than four years. Installation by even the most klutsy home mechanic shouldn't take more than three-quarters of an hour.

Single-handle mixing and mono faucets are a drag! They cost a bit more than the two-handle variety. They look pretty good, and they're mechanically quite simple. But some of them, and I don't care to tell you which they are to keep from being sued, are damn near impossible to repair. The best of them is the top-of-the-line Delta, but it costs an arm and a leg. Delta imitators are producing products which resemble the real thing, but many of them just don't cut it. They are made of cheap materials which have a tendency to fail. There are even more expensive single-handle mixing faucets on the market than Delta, but frankly, I don't think the additional investment is warranted. Remember, these are opinions. You and other professionals may have another view.

If you do buy a single-handle mixer, however, take my advice. Get the type with lever handles rather than those that have the large plastic-crystal pull-up handles. Some of those are difficult for frail people to operate.

The main reason I don't like the single-handles is because many of them are installed by plumbers and handypersons who don't bother to read the instructions that come with them. That can be dangerous!

In the United States, and, I think, throughout most of the civilized world (and some places that I don't think are so civilized), the hot-water tap is *always* on the left, and the cold water tap is *always* on the right. Remember that! It's important.

It is also important to realize that some of the mono faucets (single-handle mixers) are made with their water-supply connection tubes reversed. Some of these folks have been nice enough to mark the cold-water inlet for those who are either too lazy or too stupid to read the plainly written instructions in the brochure enclosed in the faucet package. As a consequence, many of these faucets are hooked up backwards, with the cold on the left and the hot on the right.

Now, use your imagination. Suppose there were a blind person in your family who used the sense of touch to make up for loss of sight. Suppose you have a careless teenager who seems to walk around in a perpetual fog. Just suppose that you are taking care of very young children or the elderly. Add to these "supposes" a water heater set at its highest setting, which would send 160°-170°F water out the tap, and add to that the possibility that water has recently been drawn from the hot tap, by a person who is completely aware of the reverse tap arrangement. It is possible—even likely—for an unaware, disabled, or preoccupied person to turn on the hot water, thinking it's cold, and get a second- or even third-degree burn. We are taught by custom and usage that the right is cold and the left is hot. I dislike some single-handle mixers (not all—some)! And I prefer not to take chances with people's skins in my book. So, I've issued the warning.

Repairing single-handle mixers is as simple as going to your plumbing or building materials supply house. Every faucet of this type is backed up by a repair-kit package. Simply purchase the repair kit, and replace each piece. The most important thing to remember is the sequence of parts removed from the various assemblies. To this day, with mixers with which I haven't had much experience, I list off the parts as I remove them, literally describing the places from which they come, in the exact order of removal. In that way, I never make a reassembly mistake. It is a help that some of the repair kits have illustrated instructions. Again, the easiest repair kits to locate are usually Delta, or one of the Delta look-alikes.

FAUCET REPLACEMENT

Your last resort, if all repair techniques fail, is complete replacement.

As usual, there are rules which normally apply: the distance between tap inlets on kitchen sink mixers is 8″, lavatory inlets are 4″ apart, and the distance between valve inlets on shower mixers is 8″. When you remove the deck-mounted faucets from a kitchen sink, you notice that, in most cases, there is a center tapping between the main tappings. It's a fairly modern addition, intended to receive the hose of the dish sprayer, which normally rests in a fourth tapping to the right side of the right main tapping or perforation. That right hole can be used for the air-gap fitting for a dishwashing machine, and there might be a fifth hole to the left of the left main tapping, intended to receive either the air-gap fitting or the sprayer.

The varieties of kitchen-sink centersets are many and confusing. Some of them have extended tubes beneath their bases. These tubes usually go through the center tapping on the sink deck. Other faucets have extended tubes that go through the normal right and left tappings at 8″. Some add a tube or smaller threaded barrel for the dish sprayer, which usually extends down from beneath the swing spout, through the center tapping. Others, whose main hot and cold inlets extend through the center tapping on tubes, have left or right tube or barrel for the sprayer.

Important note: Before you install your new mixer faucet, read the manufacturer's instructions carefully. *Don't take a single thing for granted. Don't listen to your neighbor, who "knows" because she or he has already installed the same type or model. Read the instructions yourself!* They may have done it wrong. If there is a question in your mind, take the instructions in hand, return to your supply store, and ask questions. If you are not satisfied with the answers, postpone your installation until you've had a chance to write to and receive answers from the manufacturer. You will regret hasty or all-advised action! Believe me.

CAUTION: *Some homes have had their plumbing systems renovated by genuine idiots, out for a fast buck. These predatory creeps know how to join pipe, but often have too little knowledge and make mistakes.*

I've been into dozens of homes in my long career, where right for hot and left for cold were whims of convenience. At the time of the construction, obviously it was easier to bring the hot line up on the right and the cold on the left. Then, rather than taking the time to cross them in the walls to orient them properly for discharge into the fixture, they've simply crossed the supply tubing from the angle stops to the faucets. I've already told you how dangerous that can be for the disabled or feeble. So, now, I'll ask you to be sure, when you've made your connections, especially if you've installed a single-handle mixer, that left is hot and right is cold. If it isn't, then simply cross your chrome supply tubing beneath the sink, and be content with a somewhat "Mickey Mouse" installation, but one which is correct and safe, in terms of the final result.

If, for some reason, you are forced by circumstances to go along with an incorrect installation, you MUST make sure that the controls are plainly marked and that people are warned before they use the fixture. An example is the reverse-plumbed shower. There's simply no way to manipulate the pipes in the wall, without tearing it out. So, make sure that the right handle is plainly marked hot and the left, cold. Don't let unaware, disabled, or weak persons use the facility without warning.

REPLACING THE KITCHEN-SINK CENTERSET AND OTHER FAUCETS

The biggest chore is getting the old centerset out. The majority of the older mixers are clamped onto the sink deck with hex nuts around the inlet barrels. To get them off, you must first disconnect the water supplies completely from the faucet. Before you do that, make measurements, and go to the store to get new supply tubing for the installation between the angle stops and the faucet inlets. You usually can't save the existing tubing—you may, with extreme care, but the time consumed could be excessive, and prove to be a waste in the end.

Use your eyes and brain now! Look at your old faucet inlets, beneath the sink. If need be, use a flashlight and mirror, so that you don't have to crawl underneath from the very start.

If the old faucet has extended tubes to which the supply tubing is attached, and your new faucet has shorter threaded barrels, it is obvious, from the start, that the present supply tubes may be too short for your new faucet, unless they are very deeply looped and have a great deal of slack. Even then, the act of straightening out the tubing could cause difficulty with crimping, especially if the tubing is the old-fashioned, heavy-wall, smooth, chromed variety. If the tubing is the very flexible modern type which appears to be made up of accordian-pleated chrome segments, you may be able to salvage and reuse it with your new faucet. Anyway, it's worth a try.

There are two sizes of chrome supply-tubing, ⅜″ and ½″. The most commonly used is the ⅜″, but the deciding factor is the size of the outlet of the angle stop beneath the sink. It's easy to determine the proper size. Simply unscrew the small nut on the outlet of your angle valve, remove the tubing with the nut, and measure the tubing size. If you'd rather make completely sure, take the tubing with you to your supply house and make a visual comparison with the products on display.

The prepared chrome-tubing connectors come in the two popular sizes and in lengths of 12″ to 24″, in 6″ increments. There is a special 9″ length for water closets (toilets). Tubing connectors are usually marketed in pairs, but the special toilet connectors are sold as singles. It is best to get the special single connectors for toilets because they terminate in fittings designed for that application only.

If you are going to discard the old chrome tubing, and it is fastened into the angle-stop nut with a small brass ferrule band, you are going to have to get a new ferrule band for your new tubing, since you will probably have to cut the old tubing to remove the angle-stop nut. Whether or not you are able to remove the nut without cutting the tubing, get a new ferrule band, since the old one cannot be removed from the tubing, and you must have a ferrule band to make a watertight joint.

If the chrome tubing is joined to the angle stop with a rubber or composition cone washer, which is pressed against the socket in the angle stop and the tubing by the nut, then, when you remove the tubing, you must also replace the cone washer. If you are unable to buy a new cone washer for your angle stop, you must replace your angle stop with the more modern variety, of the brass ferrule-band type. To do this, of course, you must turn off your house water at the main.

If, through a stroke of good fortune, you are able to salvage your old chrome-tubing connectors, you will notice that the larger nuts at the top of the tubing, which join to the faucet connections, cannot be removed. Don't let that worry you because, 99 percent of the time, they will fit your new faucet connections. Those connections are generally interchangeable. The ⅜″ or ½″ nuts on the angle stops are standard as well, and replacements can be purchased in any store that sells plumbing products.

Obviously, you must loosen the connector nuts from the water supplies to be able to remove the nuts holding the centerset in place. If the old faucet has been installed within the last ten years, chances are that the connectors are just a bit tighter than hand-tight, and should come off with a good tug. Use your basin wrench, if it is too difficult to get another type up into the recess. Look at the basin wrench before you use it. Manipulate the jaws, and the principle of its operation will become rather apparent.

When you begin to work beneath the sink, I suggest that you clear out the entire undersink cabinet of everything. If the cabinet has a very narrow opening, and you are extra large, you may have to work on the job by feeling your way along. An alternative is to get a good sturdy mirror and a work light, and set them on the floor in the back of the cabinet, with a prop under the mirror, if it needs an angle to permit you to see into the work area.

If the cabinet is both narrow and deep, and your arms are too short to reach into the critical work areas, you may have to remove the sink from the cabinet, by whatever means you are able to devise.

Usually, however, there is sufficient room for an average, husky man to work, lying on his back, with his head right under the work area. If the cabinet floor is a few inches higher than the kitchen floor, I suggest that you build a "platform" out from the cabinet lip, or at least pad it out with newspapers or rolled-up material of some sort, so that your back doesn't lie over the sharp edge of the cabinet. A slip, while straining, or someone falling over you, while you are bridging that sharp edge could seriously sprain or even break your back. It is best to wear safety goggles while working on any job which requires you to look up at the work. After all, you don't want dirt particles falling into your eyes. Falling nuts and washers are no picnic either.

You may find that the water-supply connector nuts are "frozen," especially if they were put on over ten years ago. It used to be that plumbers' apprentices did most of the finish-plumbing hookup, after fixtures were placed. They, and even some experienced journeymen, tended to overtighten supply fittings. Years of hot-water expansion and contraction and the nature of the older materials tended to weld them to the surfaces against which they were compressed. Some plumbers used such compounds as litharge-and-glycerine as well as other hard-setting pipe dopes to insure leak-free connections. Some of them are nearly impossible to undo.

After you've tried everything else, the last resort is heat. You have a propane torch in your kit. Now's the time to crank it up, and apply heat to the connections which stubbornly refuse to be wrenched out. The principle of heat application for purposes of loosening "frozen" connections is very old in plumbing. You must apply the heat directly on the nut and try, as best you can, to keep the flame from hitting the male thread or threaded barrel upon which the nut is frozen. This should cause the nut to expand and break the seal between it and the male thread. If you heat both the nut and the male thread around it at the same rate, they will both expand equally and that separation may not take place. Even so, the heat alone should help some, whether both surfaces expand at the same or different rates. Nine times out of ten, if you can wrench the nut while it's still hot, you should be able to get it started unscrewing.

I don't know how to say this without sounding condescending, but I must, since I recently had cause to observe a do-it-yourself disaster in the making. *Do not heat plastic nuts!* The resulting goo is damn near impossible to get off the fittings.

Yes, some faucets and supply-tubing assemblies come with plastic nuts and other plastic parts, such as washers and gaskets. Don't get near them with your propane torch. The stink is bloody awful.

If all goes well, by this time you have your old centerset off and you've scraped the old putty and built-up grime from the sink deck. Make sure that the mounting area for the new faucet is squeaky clean.

Most new-fangled faucets come with rubber or plastic gaskets which fit up underneath the faucet base. These gaskets are intended to make a watertight seal between the centerset and the sink deck. They will, if they're installed properly. Notice the configuration of the gasket. When you put it in place make sure that it has the correct side up and forward. If the base of the centerset is rounded in front and flat in back, the gasket should have a bead around the edge. Some instructions call for the bead to be up, to fit into the base of the faucet, and others instruct you to put it down as a part of the watertight seal. Follow the instructions!

Some faucets fasten in place by hex nuts tightened over the inlet barrels, and others have threaded rods which come through the sink-deck tappings and are cinched up by small nuts which hold cross-bars or large metal discs in place. Again, follow the instructions!

If you have the type of faucet that doesn't have a plastic sealing plate, then you're going to have to seal the centerset to the sink deck

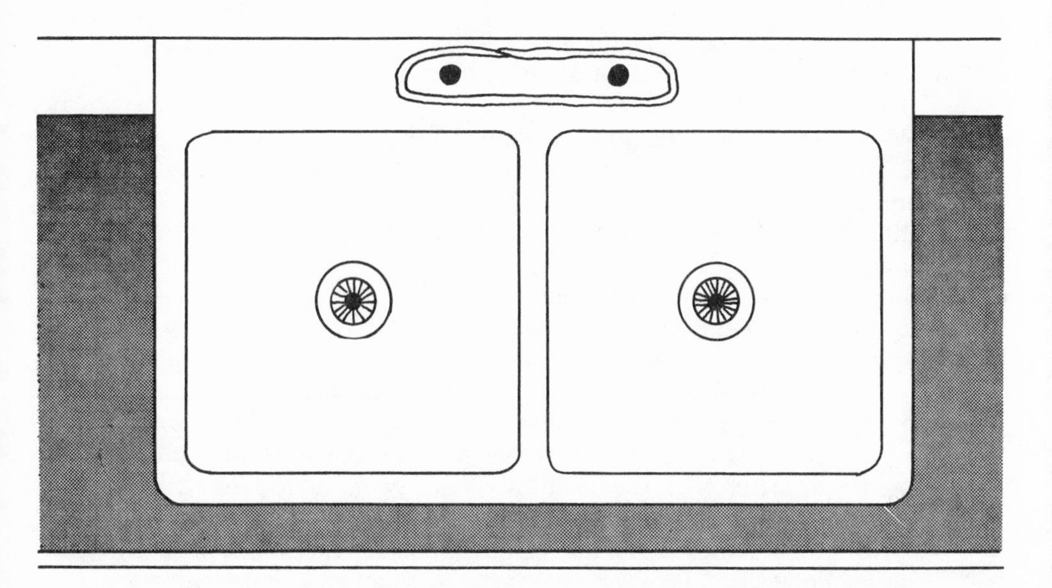

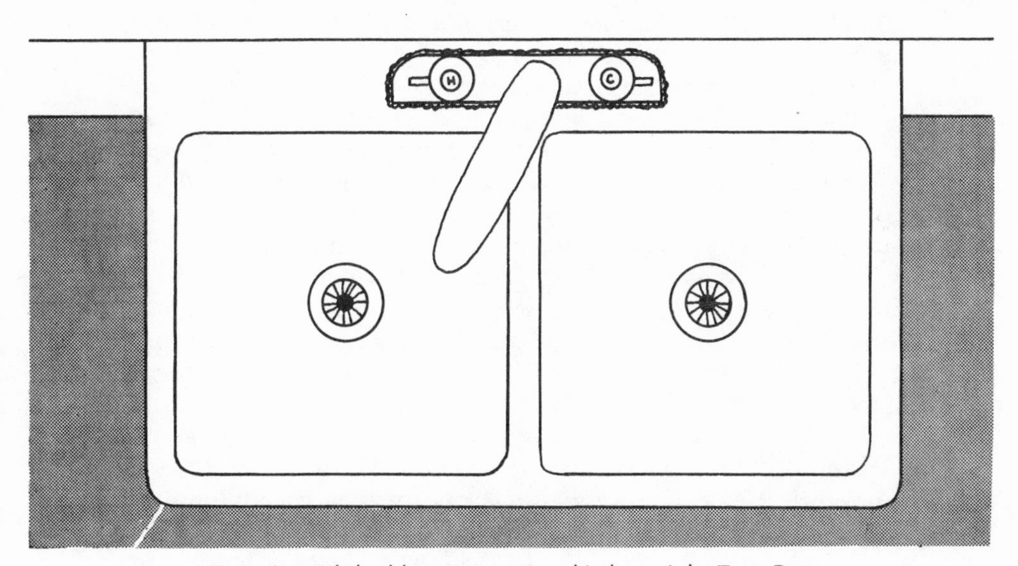

Top view of double-compartment kitchen sink. *Top:* Putty, rolled and in place to receive faucet; *bottom:* Putty squished out after faucet has been set.

with plumber's putty. The best way to do that is to take the putty out of the container and form a ball about 2½" in diameter. Then start rolling the ball between the palms of your hands. It will begin to form a rope-like mass, which you can continue to roll out letting it fall to the counter as you go. When you've finished your kneading and rolling, you should have a rope of putty over a foot long. Take the putty rope and place it on the sink deck in about the configuration of the faucet you are going to install. Then, carefully, bring down the base of the mixer faucet and make sure that you gather the putty under the base with the ends of your fingers as you press it home. Now, tighten the centerset in place from underneath.

The putty should squish out from the base all around, to be sure that you have a good watertight seal. If you've missed a spot, do it over until you have it right, or you're going to have water dripping down into the undersink cabinet from that day on, rotting the wood and stagnating in pools in the back of your cabinet.

Now to button up the job. If you are installing new chrome-plated supply connectors between the angle stops and the new faucet inlets, first put the angle-stop compression nut on the bottom of a tube. Then, slide the brass ferrule band over the end of the tube and up into the nut. Stick the tube into the socket of the angle stop, holding it firmly in place as you bring down the nut with its ferrule band and begin to screw it into place. In most cases, it won't start easily, because the new ferrule band is uncompressed and will resist going into the socket. That's normal. Just press on the nut harder, making sure that the supply tube is *straight* in the socket. If it goes in at even the slightest angle, the nut probably won't grab, and you'll be turning-and-cussing, and turning-and-cussing, and, finally, blowing it! Patience—the tube *must* go straight into the angle-stop socket. That means that, in most cases, the bottom of the tube goes in place before the top. The top is easier, because the nut is bigger and the joint more flexible.

HELPFUL HINT (super important): The tendency is to overtighten these small angle-stop nuts. Please don't do that, or you'll create a leak which will be almost impossible to correct without the use of cheaters like silicone adhesive or epoxy filler, which can mean later problems when you go to make another replacement, or remodel in a few years.

Tighten the nut until you feel it begin to bind a bit as it pushes the ferrule band into the angle-stop socket. Do the same with the other connecting supply tube, and then fasten them to the faucet connections, making sure that their dome washers are in place. CAUTION: *If the chrome supply-tubing connectors come equipped with thin brass washers, quite often they are secured in the package between the dome washer and the upper bead or shoulder of the connector tubing. Those flat washers are designed to be slipped over the supply tubing* before you put the angle-stop nut in place. *The washer should fit up under the bead or shoulder, and the nut that connects to the faucet barrel or tube goes on after the brass washer, to give solid support to the nut as it tightens the dome washer up against the seat on the barrel. If you leave the flat brass washer in place between the rubber dome washer and the bead, and the hole in the connector nut is so large that the bead will pass through that hole, then the nut will only tighten the dome washer against the barrel. The tubing itself will be loose, and will leak, eventually blow out, and flood your kitchen. If you happen to be on vacation when that happens, you'll have a houseboat when you return, only the sea will be on the inside.*

The top nuts should not be tightened more than hand-tight plus a quarter turn with a wrench. Frankly, even the wrench is unnecessary. The cone washer seals quite adequately when the nut is hand-tight.

After both ends of the supply connector are secure on both the hot and cold sides, open the faucet and then both angle stops. Close the faucet and observe the fittings you've just made. If there is no leak, fine. If you see "weeping" at any connection, turn the nuts one face at a time (less than a quarter turn), until the weeping stops. After each turn wipe the nut or blow on it to remove the water, so that you can see if the leak is being controlled by your tightening.

You shouldn't have much of a problem with the top nuts. They're larger and they press against rubber or neoprene, which usually seals quite easily. The two bottom nuts going into the angle valves are the most likely to give you problems. The cause is simple. The modern, accordian-pleated, very flexible supply tubing is a boon to the plumber and mankind, but its walls are thin; when the nut compresses the ferrule band on it, the tubing gives and bends, and does not allow the ferrule band to bite into the surface and create a solid dam against the water. The bending allows the water to go beneath the ferrule band in very small quantities, and you have weeping.

The answer to this problem is to turn the nut to the place where the weeping is nearly gone, and then let the system "shake itself out" overnight or for the next few days.

As with testing drains, run the hot water by itself for a minute or so, then the cold, and finally mix. The expansion and contraction will accelerate the "shakeout."

Often, when you return to make your final adjustments, you will find that the leaks and "weeping" have "cured" themselves. If they have not, then continue taking up the angle-stop nuts a face at a time.

HELPFUL HINTS: If you fail to control the seeping water, you have two choices: Get a new tube and ferrule band and start over, or shut off the angle stop and undo the nut. Put a small dab of clear or white silicone-seal caulk or adhesive on the end of your finger. Lifting up gently on the tube so that the ferrule band comes up out of its seat a bit, coat the underside of the ferrule band to the tube itself, so that you seal the hairline between the ferrule band and the tube, and let the tube and ferrule band spring back into their place in the angle stop. Then coat the top of the ferrule band lightly, again to the place where it joins the tube, and retighten the angle-stop nut. Allow the side upon which you've worked to stand without pressure for twenty-four hours, with the angle stop closed and the faucet open, so that even if the angle stop doesn't quite shut down the line, the pressure is relieved.

The sealing method I've just described is the most effective and least frustrating to "break" if and when you want to work on the angle valves or faucet later on.

At this point, I would be remiss if I didn't make you aware of a new plastic product on the market, which I have used only when my customers have requested it. It's not that I don't trust plastic. God knows that it is used in almost every durable product manufactured today. But I have always hesitated to use plastic in hot-water applications. Yet, the manufacturers assure us that their products are quite utilitarian and durable. Even though they are relatively new items, you may find the flexible, glass-strand-reinforced plastic water connectors very good for your connections to sinks, lavatories and water closets. They are equipped with swivel nuts of precisely the correct size to make on to the angle stop and faucet water inlet barrels or tubes. They don't require ferrule bands or cone washers. Everything is built into the tube connections—talk about convenient!

Remember, they violate the 1979 *UPC*, which doesn't permit the use of plastic pipes and tubing for cold- or hot-pressure water applications in a house or building. The only exceptions I can think of are items such as hand-held shower heads, dish sprayers, and the like, because they are considered fixtures or equipment and don't constitute a permanent portion of the house plumbing system. I have been informed that the new code will probably modify that stand.

In any case, feel free to use these new plastic supply connectors, but make sure that you don't get such short ones that you have to strain them to make your connections. It would be better to have slack, because plastic has a hell of a modulus of expansion, which means it expands and contracts over a much greater range than metal. So, if you install your tubing on a hot day, and it's exactly the correct length (within the quarter inch) or even taut, then when you run cold water in it or the weather turns chilly, the plastic will contract and put strain on its coupling nuts. It is said that the glass-strand reinforcement should keep them tight, even under these circumstances, but, realizing the nature of the materials with which we're dealing, I say, don't take the chance. Give the plastic tubes plenty of slack.

The first products that were marketed had a tendency to leak at the lower nut (the one that goes onto the angle stop). I am told that the new supply tubes are sound, and that the problem has been eliminated. I hope so.

The question now comes: What do you do if you want to upgrade your kitchen sink at the same time as you install your new centerset? You want to install a dish sprayer, "pig" (garbage disposer) and dishwasher. Fine, we'll talk about that right now, rather than referring you to the section on construction (Section Five).

INSTALLING YOUR NEW DISH SPRAYER

Obviously, you're going to have to buy a centerset which includes the dish sprayer in its design. Before you buy it, however, take a good look at your sink. The dish sprayer has to be installed somewhere. If the sink is a three-holer, you're going to have to buy the type of centerset that has all of its tubes (hot and cold, plus dish-sprayer tapping) coming out the center of the base. This type usually has the sprayer mounting right in the base, and the hose for it comes up the right tapping in the sink. The base is usually secured to the left sink tapping by means of a threaded rod, bridge piece, and nut. Also, these types of faucets are always single-handle swing-spout centersets.

Other sinks have a fourth tapping to the right of the centerset tappings. Occasionally, you may find the extra hole on the left of the hot-water tapping. If it is not in use, this extra hole is usually covered with a chrome, stainless steel, or plastic cover. With a fourth tapping, you can use any type of sprayer-equipped centerset, unless, of course, you plan to install your new dishwasher at the same time. You have to have a place to put the air-gap assembly from the dishwasher. It is usually installed on the sink deck. If you plan a dishwasher, you're going to need that fourth tapping, and will require the single-handle faucet I've described in the previous paragraphs.

If you're not going to install a dishwasher, *ever*, go right ahead and use the fourth tapping for your sprayer.

Of course, you can always consider the installation of a new sink with five tappings. That will take care of everything in one fell swoop.

You may have to install a new sink anyway, if you want to enjoy the convenience of the garbage disposer. Some very old sinks have drain holes which are too small to receive the garbage-disposer mounting flange. In that case, your job is cut out for you.

HELPFUL HINT: As soon as you start hanging expensive "jewelry" on your sink, you have to begin to think about where things are going to go. Here's a rule of thumb: If you use a unitized single-handle center-supplied kitchen swing-spout centerset, a four-hole sink deck is all you'll ever need to get you into the twenty-first century. If you use a two-handle centerset, and you want a dish sprayer and dishwashing machine, you will need a five-hole sink, or you will have to drill a cabinet, which is not recommended.

CAUTION: *Don't even think about installing your dishwasher without an air-gap assembly, as so many people have done. I'll explain why, when we get to the installation of the dishwasher.*

Okay, we're ready to get on with the sprayer installation. *Read the manufacturer's instructions.* If you haven't a tapping in the sink deck for the sprayer, of course, you've had to buy the single-handle "special" I've described.

HELPFUL HINT: If you have a stainless-steel sink, and you're handy with sheet metal, perhaps you can drill or cut out the extra hole. It's impossible with enamelware or ceramic, of course.

The instructions that come with the faucet are usually sufficiently clear to make a proper installation. The important thing to remember is that the hose connection at both ends must be secure. Usually, it is not necessary to use any kind of pipe-joint compound or "dope" on these threads. They are commonly brass-to-brass connections. If they are plastic to metal or base-metal to brass, perhaps a bit of pipe-joint compound is in order. If you are sealing plastic, don't use the type of compound which has a volatile solvent base (smells like acetone, or airplane cement), because the solvent in the compound will dissolve the plastic, and you'll be out of luck.

After you've installed your sprayer, check its mechanical function, flip the lever down and up several times. If the main water flow to the swing spout is not almost completely shut off, and the spray stream strong, the little automatic diverter in the faucet mechanism may have some pipe compound or particles lodged in it. That may require faucet disassembly and a clean-out. If you're not into that, just take the faucet out and return it. It could be defective. Most stores will simply replace it.

For you who have wall-mounted kitchen-sink faucets, the job of replacing the unit is relatively easy. Simply turn off the main house water supply, open both sides of the faucet to drain it, open a faucet or hose bibb lower down than the sink faucet to get the pipes reasonably clear of water, and loosen the two union nuts that attach the faucet to the wall. The union nuts will usually have gaskets between faces, to seal the system. Some faucets have ground-joint precision fits and are without gaskets.

Unscrewing the unions will allow the faucet body to be removed. All that remains is to unscrew the mounting portions of the unions, and replace them with the new assemblies. Since the universal distance is 8″, wall-mounted kitchen faucets and laundry-tub swing-spout mixers are interchangeable. In very old houses, you may find 4″ and 6″ centers, in which case you must buy the type of faucet which has a built-in adjustment to the proper distance. Those are usually made in rough-brass natural or chrome finish, are not very attractive, and normally would be used to supply laundry tubs only. But they can be used in the kitchen.

Dish sprayers are often mounted directly on the wall-mounted kitchen swing-spout faucets, and attachments are available for some of the older models, but when it comes right down to it, the most esthetically pleasing types of kitchen-sink faucets are the deck-mounted models. To convert from wall to deck mounting is a major plumbing operation, which I shall deal with in Section Five.

DISHWASHER INSTALLATION

You are normally prevented from installing a built-in type of dishwashing machine unless you have a deck-mounted centerset. You can use one of the portable types, however. Those units are commonly mounted on casters, and have hoses for water supply and waste-water discharge that attach to the wall-mounted faucet and deliver into the sink bowl.

It is best to install either type of dishwasher in an undercounter recess. This can mean giving up drawer and/or cabinet space, but, if you want a dishwasher, you must be prepared to make the sacrifice.

HELPFUL HINTS: Before you even buy your new dishwasher, have the seller go over the installation instructions with you, in detail. *Read the instructions* yourself, right there in the store, and, if you have any questions, ask them then. Most important, list off the height-width-and-depth recommended-mounting dimensions. Take note whether those dimensions make allowance for the tubing, hose, and electrical items which must be installed with enough slack to permit the washer to be moved out from under the counter for service and repair. I have often found that the manufacturers' dimensions haven't taken those important factors into consideration. *It is far better to leave extra room and face your cabinet with a six-inch wide panel between the undersink cabinet and your washer cabinet, so that you can service the machine easily.* Often that is next to impossible, because the washers are usually 24″ wide, 35″ high, and designed to be installed in cabinets 24″ deep. The only way you'll get your extra 6″ is if you have a 30″ cabinet next to the undersink cabinet. This implies some revamping, if you want to do a really top-notch job. Otherwise, you'll have to go with what you have. Most machines have enough space behind the working mechanisms to allow for the coiling of extra hose, tubing and wire, but believe me, it's a tight fit.

MORE HELPFUL HINTS: Many plumbing supply houses stock a regular dishwasher installation kit complete with tubing, hoses, and fittings. These are designed to make hook-up to existing plumbing quick and easy. CAUTION: *Many of these kits do not meet code requirements. Some of them have galvanized tees which must be installed between copper plumbing and your brass angle stop, plus an accordian-type water-supply tube, 24" to 36" long, which is intended to be connected from that tee to your new machine. First of all, the galvanized tee, combined with the copper and brass, will insure galvanic corrosion, and rapid deterioration and clogging of the tee. Second, that accordian-type tube is usually non-certified for the service for which it is specified. Its walls are often too thin for its diameter. With that galvanized tee in the circuit, there could be leak problems later. If you use this kit, spend the extra money to buy a ½" brass tee to replace the galvanized one, and go ahead, realizing that the hook-up is still non-code, because of the water tube. If you have a galvanized steel plumbing system you can use the tee in the kit, but use a dielectric union between it and the dishwasher water supply tube.*

A dielectric union is a type of insulating coupling which prevents most of the interaction between dissimilar metals, and usually prolongs the life of the system involved. There are also less-expensive insulating couplings available, but they are not as common as dielectric unions, so I've emphasized the union over the coupling. Actually, any device which separates the two different metals from direct contact with each other will do the job.

Do make sure that you understand the *exact* nature of your undersink plumbing before you leave for the supply house to purchase your dishwasher. Take into account how your angle stops lie in the cabinet, the type of plumbing (copper or steel), and the length of the tailpiece between the sink and the p-trap, if you don't have a garbage disposer. If you do have a "pig" (garbage disposer), then notice whether or not your unit has a tube protruding from the side of the upper section which is designed to receive the discharge from your dishwasher. Often, quality disposers have this tapping. All that's needed to be done to activate this feature is to punch out the blank where the tube joins the body of the disposer, with a screwdriver and hammer. CAUTION: *Be sure that you remove the blank from the interior of your disposer.* It won't grind metal. In fact, leaving it there could ruin your disposer when you absentmindedly fire it up to waste your next rotten banana.

Dishwasher Drain Entering Waste Disposer (2-Compartment Sink with Yoke-Type Drain)

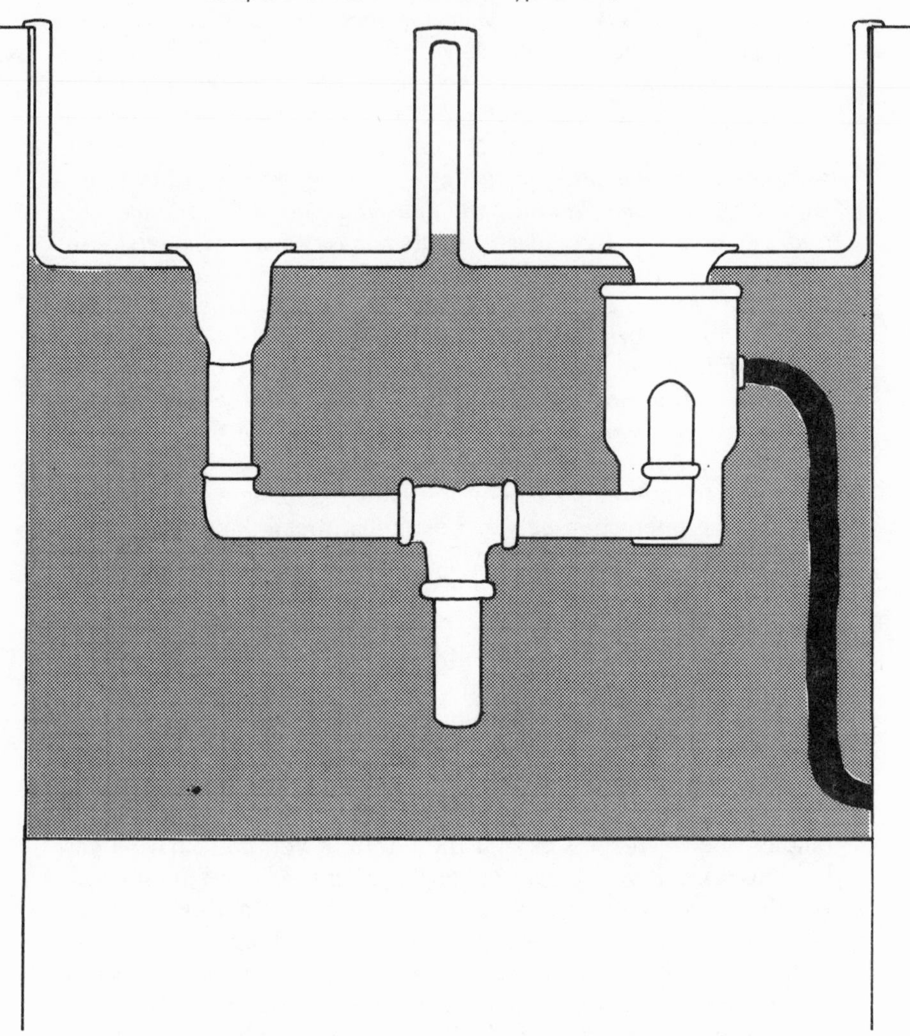

Dishwasher hose coming into the side of a waste disposer, mounted on one sink of a two-compartment unit, with drain in between the two compartments.

Air-Gap Assembly

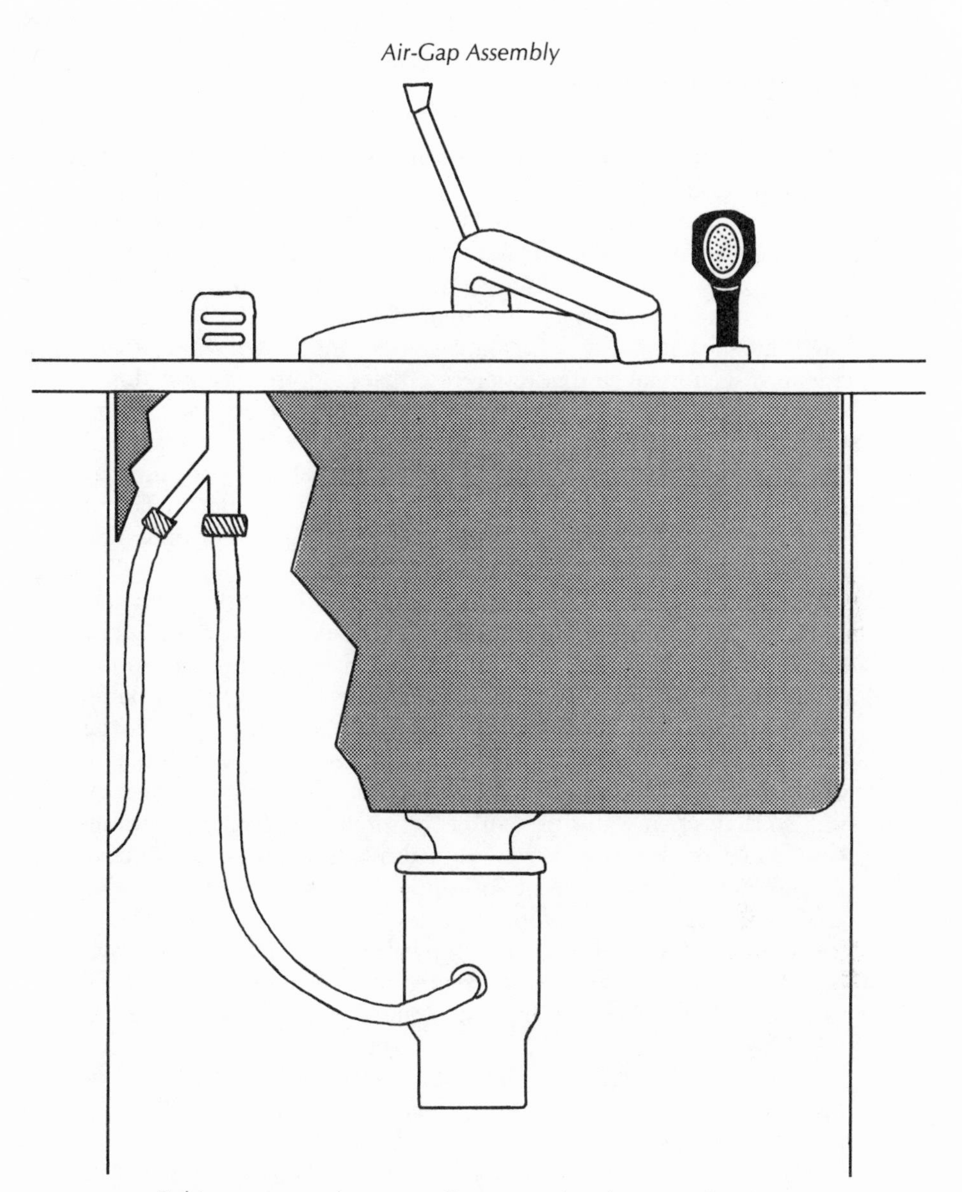

Tubing connected to waste disposer and to the vertical outlet of an "air-gap assembly" (inverted "Y" below countertop and slotted dome above it), and coming in the angled outlet.

If you don't have a disposer, you will have to discharge the waste from your dishwasher into the kitchen sink trap. This is managed through a special tailpiece, which may either be purchased separately or as a part of your kit. With or without a dishwasher installation kit, you will need to purchase an "air-gap" assembly. They are not ordinarily supplied as a component of kits.

Let me describe the appearance and function of the air-gap assembly. Its normal configuration resembles a slotted chrome dome beneath which are two plastic male hose connectors, one going straight down the other at an angle, forming a sort of upside-down "Y." These connectors are each different sizes. The rule of thumb is that the water from the washing machine comes in the smaller of the connectors and goes out to the drain from the larger one. The hose sizes are normally ⅝" coming in and ⅞" going out.

The purpose of the "air gap" is to prevent dirty dishwasher discharge water and other soiled waste from being sucked into your drinking-water lines, in the event of negative pressure or siphon action in the water lines as a result of water from the main being shut off, while a tap lower in the system is open. In that case, the water from your dishwasher could be sucked into your drinking-water lines, and contaminate the entire potable-water system with filth, so that when water pressure is restored, waste-water contaminants could be discharged from your drinking-water supply. These same air gaps, in the forms of anti-siphon valves and vacuum breakers are used elsewhere in your home's water and drainage systems. Examples are the required anti-siphon valve on all lawn-sprinkler systems, the obvious air gap that is created when you put your clothes-washer hose into the standpipe or laundry tray, and the backflow-prevention device for hose bibbs, which is code in California.

IMPORTANT CAUTION: *Do not install your new dishwasher without this air-gap assembly. Not only is it a code violation, but it is a threat to the well-being of your family. Polluted water is dangerous.*

Okay now, you're ready to plumb your new labor-saving machine to receive last night's party dishes and three dozen wine glasses. Uncrate the machine, clear the decks, and get ready to work without interruption or disturbance. As I've intimated, you've already opened up the required space, and taken your 120-volt power from its own breaker (15- or 20-Amp, which ever is specified in the manufacturer's instructions, *which you've read front-to-back*). The best sys-

Dishwasher Hose Discharging into Special Kitchen-Sink Tailpiece

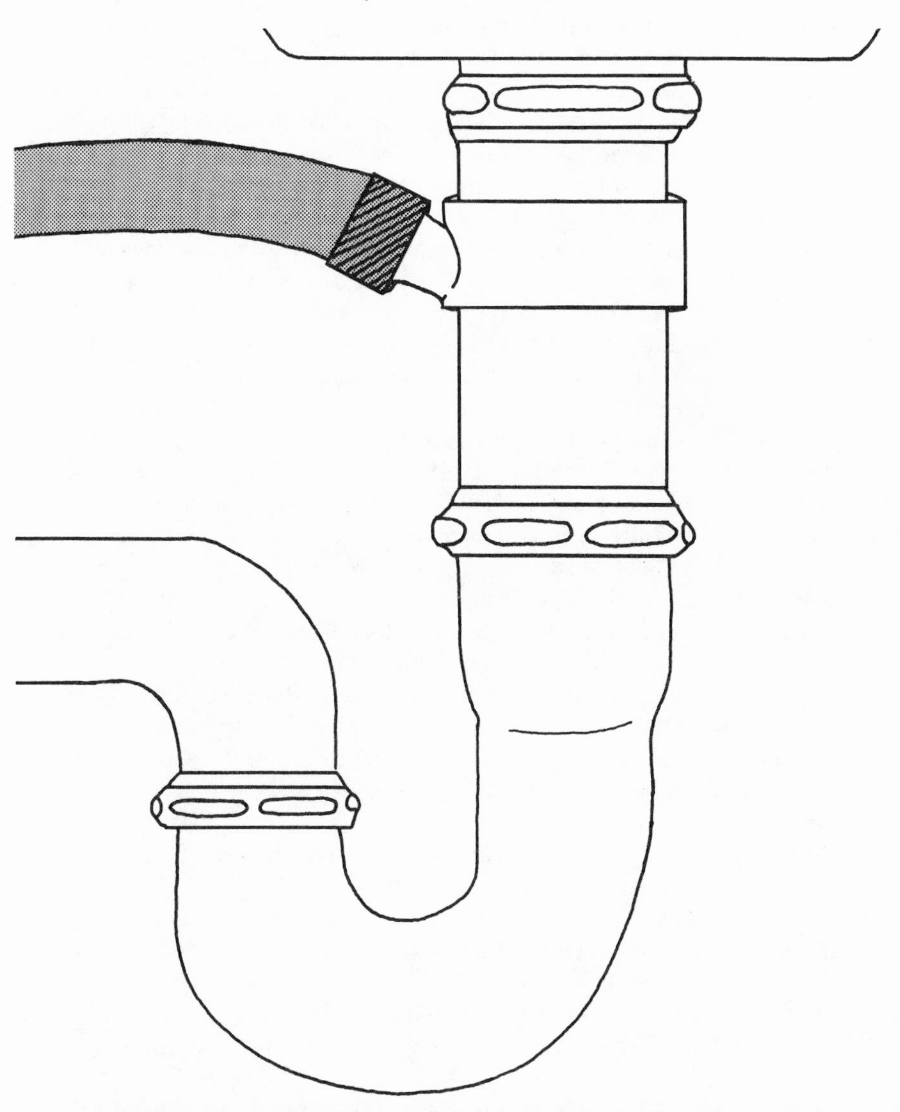

Hose coming into vertical tailpiece of kitchen-sink drain system, between two compression nuts.

tem I've yet seen terminates the power in a ground-fault-interrupted wall receptacle in the undersink cabinet to which you take your sturdy dishwasher cord and plug, through a hole drilled in the partition. CAUTION: *Be sure that you have consulted the electrical code in force in your community, before you make this important electrical connection. The ground-fault-interrupted wall receptacle is a safety device which is usually installed within five feet of any water discharge point or water-using electric appliance. It could save your life one day.*

Your first job is to prepare the sink drainage system to receive your dishwasher's waste discharge. If you have a garbage disposer, and the unit does not have a dishwasher-waste spud, you will have to improvise a discharge point into the place just before the p-trap going from the "pig" to the drain in the wall. If your "pig" has a plastic waste tube, you can purchase a plastic "saddle" for 1½" tubing, with a ½" schedule 40 outlet. If the outlet is "female" (designed to receive the end of a ½" schedule 40 PVC pipe) purchase a short length of pipe at the same time (you'll need only 3" or so). If the outlet is "male" you'll need nothing more. When you go to buy this part, take the disposer waste tube with you to be sure that the saddle radius is correct for the tubing. If it is, take it home, drill the side of the tubing, ream it to the inside diameter of the saddle outlet and weld the saddle into its place with all-purpose solvent cement, of the type which will handle PVC and ABS. If they don't have such a compound in the store, then you must get a cement for each of the materials to be joined and mix them together. Believe it or not, that system works very well indeed.

Of course, if the waste tube and saddle are made of PVC, a simple PVC solvent cement will do.

If the "pig" waste tube is metal or polypropylene, buy a brass bolt-up saddle. CAUTION: *You must take care with them because they are designed for schedule 40 pipe, so that a 1½" U-bolt saddle may be too large for your waste tube.* You may have to adapt from a 1¼" saddle, or "pad down" from the 1½" saddle with carefully crafted rubber or neoprene pads. ADDITIONAL CAUTION: *Take up on the saddle U-bolt(s) only until you achieve a seal. If you overtighten on chrome tubing you will most likely crush it.*

Finally, if your "pig" is so old-fashioned as to not have the spud inlet, often it has a discharge-tube adapter which can be purchased as an "option." In the case of the odd "pig" or two, just go to the

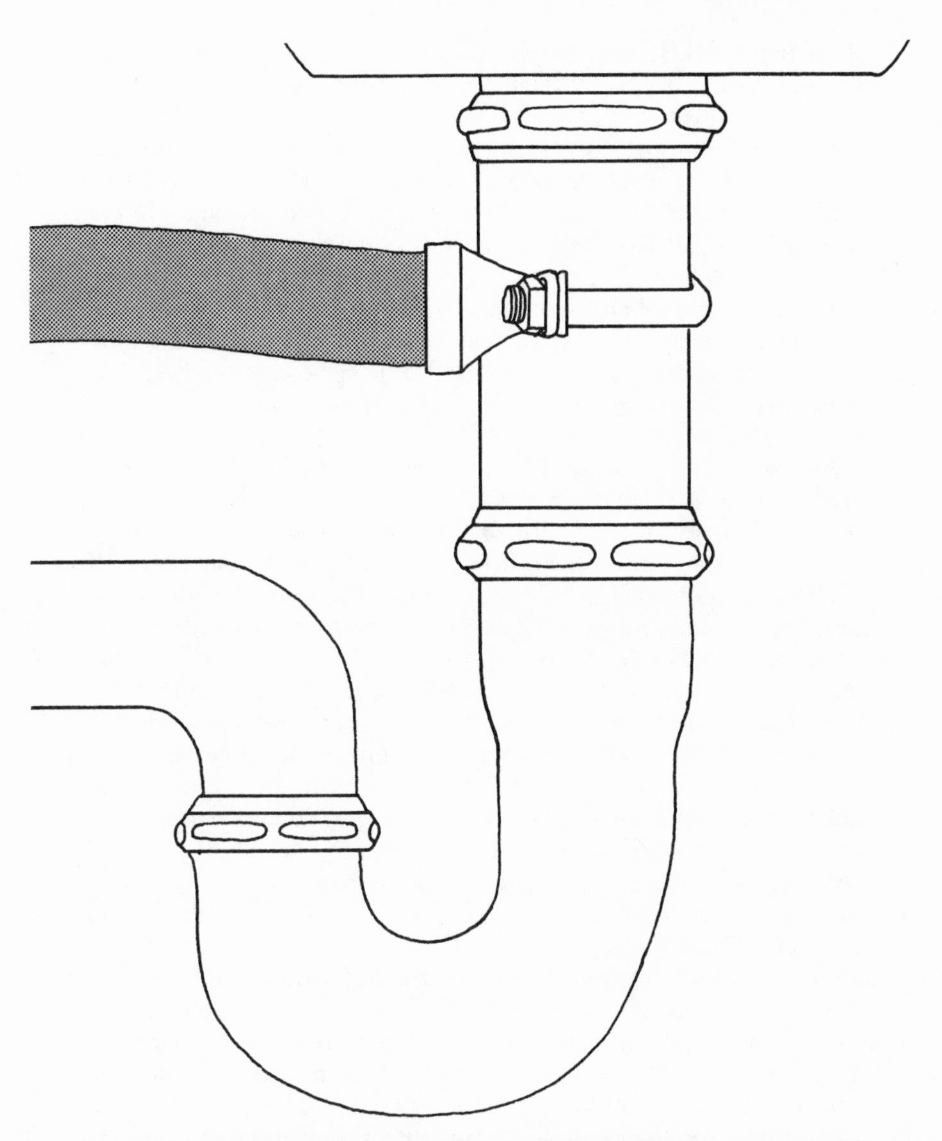

Hose entering special bolt-up saddle, attached to kitchen-sink tailpiece.

store that markets that brand and make inquiries before you go through all the trouble of adapting on your own, by means of saddles.

If you have an old disposer, perhaps it's time to replace it with a new one. All new machines come equipped with the required spud.

If your sink does not have a disposer, but rather discharges into a straight piece of tubing called a "tailpiece," all that must be done, in most cases, is to replace the existing tailpiece with the special unit mentioned earlier. For assembling and securing p-traps and the tubing going to them, use the techniques that I have described in the sections on unblocking drains (page 37). Have new gaskets and compression nuts ready in the event the present ones look grungy and cracked.

Put your "air-gap" assembly in place on the sink deck, in the appropriate hole. Make sure that you have drilled two 1½" holes in the cabinet side, between the washer and the undersink cabinet. It's usually best to drill them in the rear, if you have a tight-fitting machine. If you've left the six-inch gap between the machine and cabinet partition, you can drill your holes wherever they will be convenient.

IMPORTANT HELPFUL HINT: *Do not, repeat, do not skimp on the hose or water-supply tubing. Leave plenty of slack in both so that you can move the machine out of its cabinet for repairs and service.*

If your water-supply flexible tubing is only 24" to 36" long, install additional rigid copper or other pipe, right over to the machine cabinet, and let all the flex be in the cabinet to give you enough slack. If your flex is 48" to 60" long, you probably won't have that problem. The important thing is the slack. Follow kit instructions to the letter.

If you do not have a kit, the code and common sense will stand you in good stead.

You must have a separate valve for the water supply to your dishwasher. Normally, you tee off from the hot-water supply to your sink and use a straight stop. The straight stop is an angle stop straightened out. The water does not go in the bottom and come out the side. It goes in the bottom and comes out the top. It's a kind of globe valve which resembles an angle stop. Those made for dishwashers are normally not chrome-plated, but rather brass-colored, even if they're made of pot metal. They fasten to ½" iron or copper pipe. Normally they are of the brass ferrule-band outlet type, and they usually accept ½" tubing for the washer's water supply.

To install this straight stop, either shut off your hot-water supply at the heater, or shut off the incoming water main to the house.

HELPFUL HINT: You may substitute any standard ½" globe valve, either with a tubing adapter or from which you can run pipe into the cabinet, in the case of short flex.

With your dishwasher-water-supply valve in place in the undersink cabinet, you can now turn on the water to the house.

Your washing machine will often come with the ⅝" hose as a part of the package. Some of them are equipped with both the ⅝" and ⅞" hose, plus the "air gap." Still others have the whole ball of wax, every fitting needed to make the installation, but they are rare birds.

In any case, at this time make all of your washer connections, including the water-supply elbow or adapter on the machine, and the discharge-hose hook-up. Make sure that you've removed all of the shipping blocks and fasteners from both the outside and inside. Pay special attention to the underportion of the machine and the motor. Some machines have special plastic blocks in place to prevent damage to the motor and other fragile equipment, including the relays and other electrical items.

Gently move the machine back toward the cabinet, and, as soon as you are able, begin to feed the discharge hose through its hole in the partition.

Before you've moved the machine into the cabinet proper, hook up your water-supply tubing, and test it by opening the washer valve in the undersink cabinet and checking for leaks. After the test, close the valve once more.

Move the cabinet back farther into the recess, taking care to insure that your water and drain lines are not fouled or allowed to kink. HELPFUL HINT: At this point a bit of aid would be helpful. One should take care of the lines, as the other pushes the machine into the space. CAUTION: *Make sure your electric line is connected before you push the machine too far back.*

The washer is set when the face of its cabinet is flush with your kitchen cabinets, with the washer's door open. The closed door should bulge beyond the face of your kitchen cabinets.

Before you do anything else, level your washer.

HELPFUL HINT: Leveling can usually be accomplished by placing your torpedo level on the inside edge of the open washer door, parallel with the door's hinged edge. If your machine has leveling screws, there is access through the lower kickpanel. If it is a budget model without levelers, I have found that shims made from cedar shingles are helpful. Just take a fragment of the thinnest edge of the shingle and shove it under the low side. If your dishwasher is within a ⅛-bubble of being level, it's okay.

SPECIAL COMMENT: As a point of information, ⅛-bubble on a level is usually taken to mean that one edge of the bubble reaches beyond the level line by ⅛ of its total width, which would imply that ⅞ of its width would be inside the level band of the instrument. One-eighth-bubble over the span of 24″, is about ½″ out of level. That is usually acceptable for machines of this type.

Sometimes it's impossible to level the washer because of the lay of your house, the floor, or the cabinets. In that case, you'll just have to forgo the leveling process and live with what you have. If your cabinets are as much as an inch higher (internally) than the machine, perhaps you can run shim rails inside on the cabinet floor, with the leveling built into the difference in their heights. If you do that, however, be prepared to live with the slight angle that will be quite noticeable between your countertop and the top of the machine. Realize that, in the final analysis, most machines will operate satisfactorily out of level, but if they are very seriously out of level, water can leak out of the lower edge of the cabinet if it is opened during a cycle to add dishes.

The final button-up is to screw the tabs at the top of the washer cabinet to the countertop. Amateurs almost uniformly fail to accomplish this finishing touch. It is essential to keep the machine from moving while it is in operation. Slight movement can begin to "work" your water connections. Any walking-vibration which can be prevented is a form of "leak insurance."

Before you fire up your new washer, make the appropriate connections to your air-gap assembly. The ⅝″ hose from the waste discharge of the washer should be made on to the smaller of the serrated plastic hose connectors on the assembly, and the stainless steel "radiator-type" hose clamp securely tightened to seal the hose against the serrations or barbs. At this point you take the ⅞″ hose to the garbage disposer spud (which you've prepared by knocking out the blank), or to the special tailpiece with its waste-water inlet tube or spud. Don't forget to clamp up those ends as well.

SPECIAL COMMENT: The reason the larger hose is connected

between the air-gap assembly and the spud at the point of discharge is because the assembly must have "relief." The water comes into the assembly from the washer under high pump pressure. The "air gap" is just what its name implies, a gap between the incoming water and the outflow. The larger hose will accept the volume of water coming into the assembly under pressure, compensating with its size for the inlet pressure. CAUTION: *If you try to force ⅝" hose onto the discharge side of the air-gap assembly, and onto the waste spud at the sink drain, you will have water spewing out of the slots in the dome of the air-gap assembly. The same thing will happen if you discharge the water from the assembly to a point higher than the air-gap dome, as with a drain in a specially mounted sink that is higher than the countertop in which the assembly is installed.*

With your water connections completed, wash dishes and glasses to a fare thee well. For the first day or so, keep on the alert for water leaks, which will show up on the floor in front of the machine. Connections have been known to "shake loose" during the first days of service. Don't be chagrined or upset. Leaks during "shakedown" are perfectly normal.

WASTE- OR GARBAGE-DISPOSER INSTALLATION

Americans call them garbage disposers, waste disposers, or dispose-alls. Some Canadians and Europeans call them "pigs." Whatever they're called, they are pretty much standard in American and Canadian kitchens.

Some of them are very expensive and powerful, loaded with features and gimmicks. Others are basic, designed to gobble soft garbage and dispense it into your house sewers in a form which should not cause harm to the pipes.

Then there are the middle-of-the-road "pigs," which have good design and enough power to do a quite adequate job of work.

SPECIAL COMMENT: Buy middle-of-the-road! Don't worry about that extra layer of soundproofing. It's too expensive, in most cases.

Examine your sink. If it doesn't have a drain hole large enough to accept the disposer (4" in diameter), you'll have to buy a new sink.

Before you leave the shop from which you purchase your "pig," again, as with the dishwasher, read the instructions carefully, and discuss them with the clerk. Make sure that you have all the fittings you'll need to make a complete installation.

Waste Disposer Installed in Single-Compartment Kitchen Sink

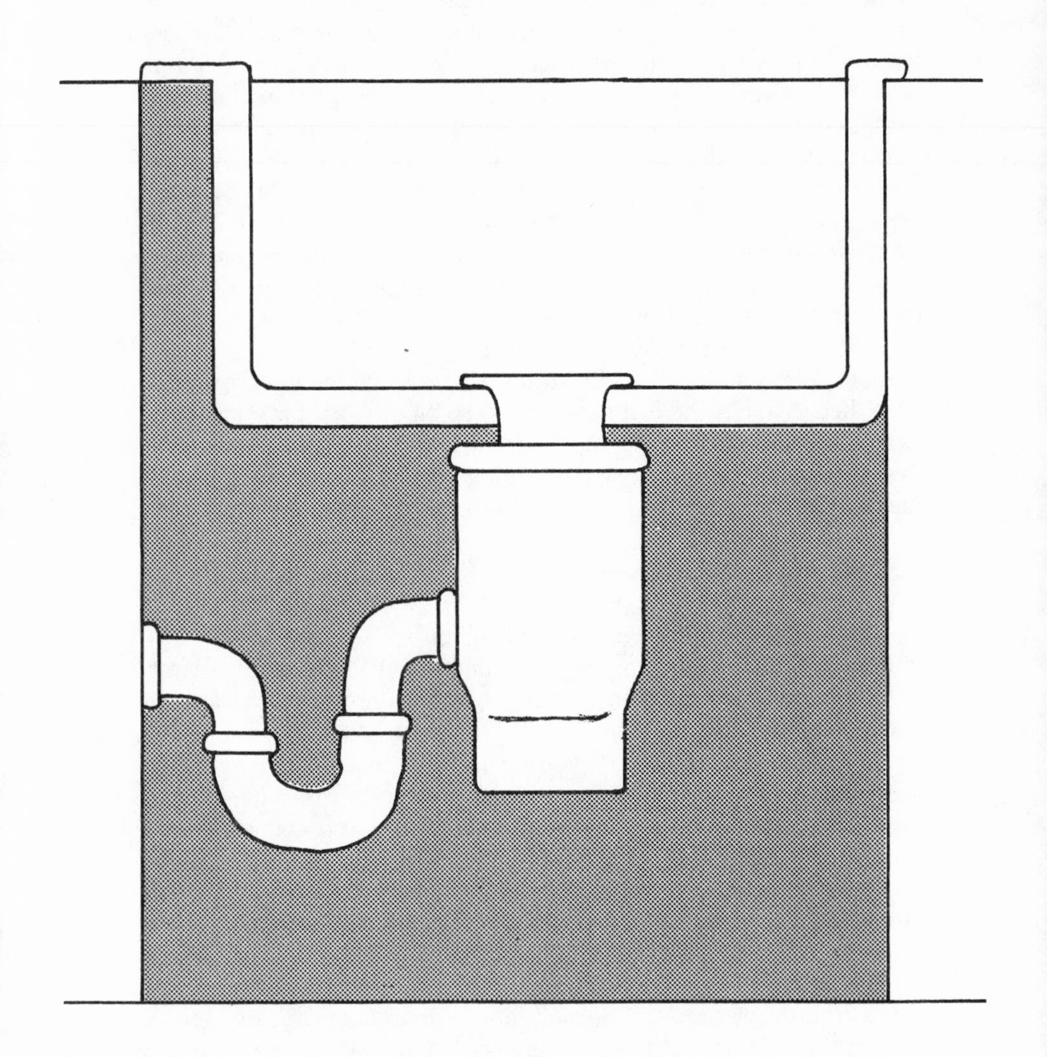

Waste disposer installed in single-compartment sink, connected to a standard p-trap.

HELPFUL HINT: It is best to figure that you're going to replace all of your undersink drains at the time of installation. Before you leave your house, memorize the drain setup. The kitchen-sink drains are almost always 1½", so you'll be purchasing new 1½" tubing and fittings. If you have a double compartment sink, it would be best to replace the drain fittings on the side that is not going to receive the disposer. Do it all at once, and you'll have many years of trouble-free service from your rehabilitated sink.

HELPFUL HINT: Almost every brand of waste disposer has its own fastening or connection system. Some of them are connected by means of pressure flanges, between which three headless bolts screw up to put a continuing pressure on special seats that are, in most cases, dimples designed to receive the sharp tips of the bolts. The lower flange is held in place with a split ring which fits into its special groove in the barrel of the flange mounting assembly. Other disposers have threaded barrels, and still others have clamping assemblies. In all cases *read the manufacturer's instructions* before you do anything.

Always have plumbers' putty handy, because you are going to have to seat the flanges of both the "pig's" and the basket-strainer's drains in putty, which is rolled in the manner described in the section on the installation of your kitchen-sink centerset.

SPECIAL COMMENT: You will have to run a 120-volt line to the undersink area. Again, you must have a separate breaker circuit for your "pig," just as for your dishwasher.

The rest of the installation follows the general guidelines I've set down for the other undersink installations. Since there is little standardization in the design and installation characteristics of the different brands of waste disposers, the manufacturer's instructions should be sufficient to get you through the job. As for the attachment of the unit to the existing drainage plumbing, that will be a matter of fitting up on the basis of trial and error.

HELPFUL HINT: While you are at the supply store, especially if you have a double-compartment sink, get some extra 1½" "extensions" complete with nuts and compression gaskets, perhaps a compression elbow, and the most appropriate double-compartment drain assembly. The easiest is the type wherein one sides goes straight down to the trap, and the other comes in on a tee. An alternative type has both sides coming to a center tee, which delivers to the p-

Waste Disposer Installed in Double-Compartment Kitchen Sink with Offset Drain (Left)

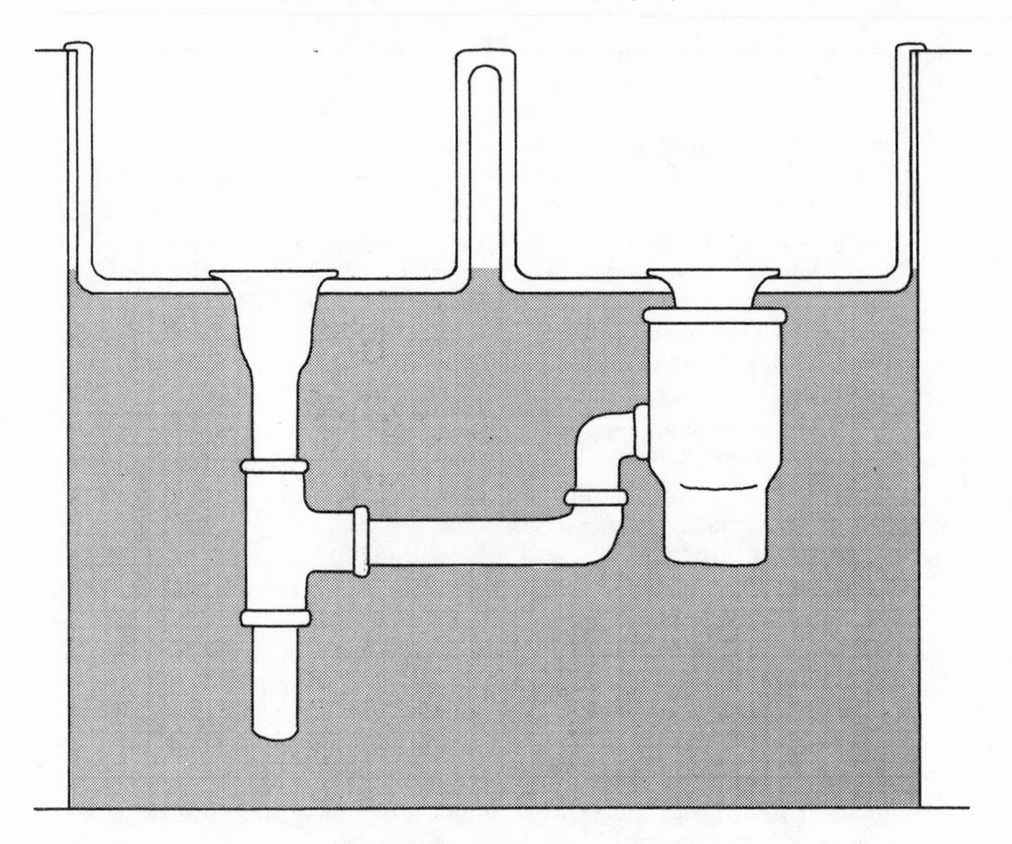

Waste disposer installed in double-compartment kitchen sink, with drain on the left.

Waste Disposer Installed in Double-Compartment Kitchen Sink with Center Drain (Yoke Drain)

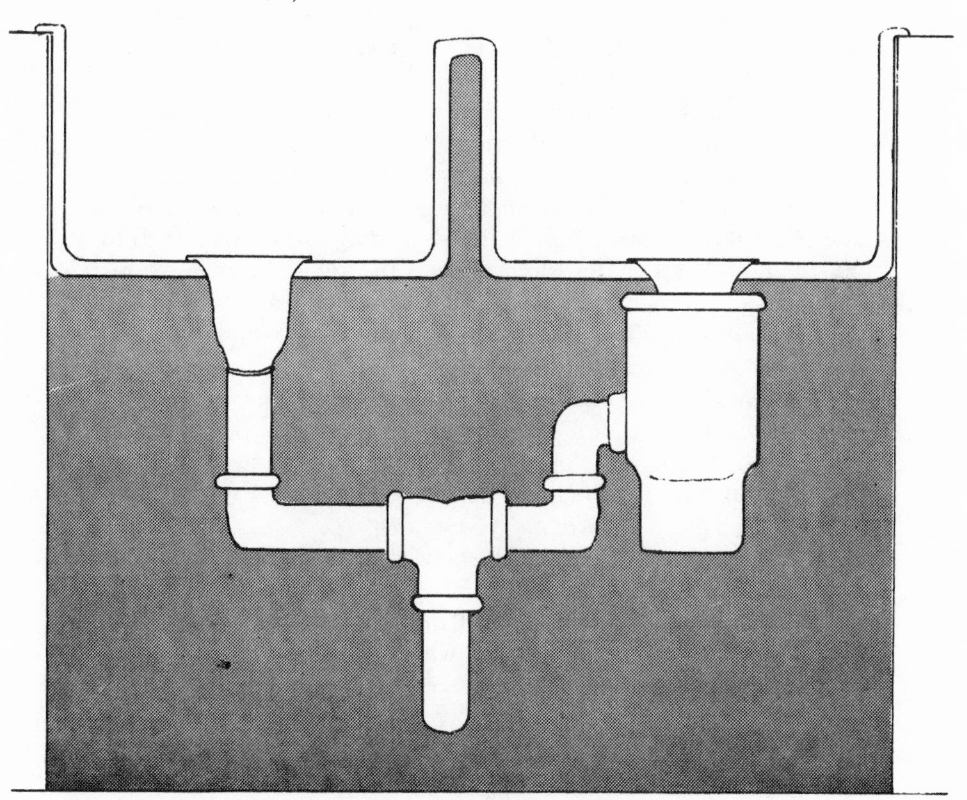

Waste disposer installed in double-compartment kitchen sink, with drain in the center.

trap. Of course, if you have a kitchen drain smack on the centerline of your cabinet, the "yoke-type" which I've just described is obviously the way to go.

If your drain is offset, the first type is best.

A lot of this fit-up depends upon the type of waste disposer you buy.

HELPFUL HINT: The disposer discharges can be swung to any convenient angle before you tighten up the machine on its mount. Take advantage of that capability.

Sometimes, you find that the drain in the wall is a bit too high, considering the level of the discharge tube from the "pig." It's easy to cut the discharge tube with a hacksaw.

HELPFUL HINT: If you must cut chrome drain tubing with a hacksaw, make sure that you build a "U"-shaped jig just large enough for the tubing to nest in the slot between the sides. Your jig should resemble an inexpensive wooden miter box. Cut a slot in it with your crosscut saw. Then place your tubing in the slot and cut it with your hacksaw. Actually, if you have a vise on your workbench, it would be best to seat the jig in its jaws to steady the working surfaces. In the absence of a vise, simply nail or screw your jig to any solid horizontal surface. You need the support of the jig, or else cutting the thin chrome tubing will be an exercise in near futility. With support, it's a piece of cake.

CAUTION: *Carefully file all hacksawed edges smooth before you attempt to fit a new piece in place.*

I make it a practice to seal the flange edge of the tailpiece that goes into the basket-strainer flange, with a thin coat of clear silicone caulk on the face of the tailpiece flange.

NEW MIXER AND BRASS FOR THE BATHROOM SINK

A lavatory (bathroom sink) faucet and "brass" replacement is certainly no more difficult than that of a kitchen centerset, and you should follow basically the same procedures except for one important item: the pop-up drain.

In most cases you will not have to replace the pop-up drain assembly. All you must be sure of is that the faucet you buy has the drill-

Lavatory Pop-Up Drain Mechanism

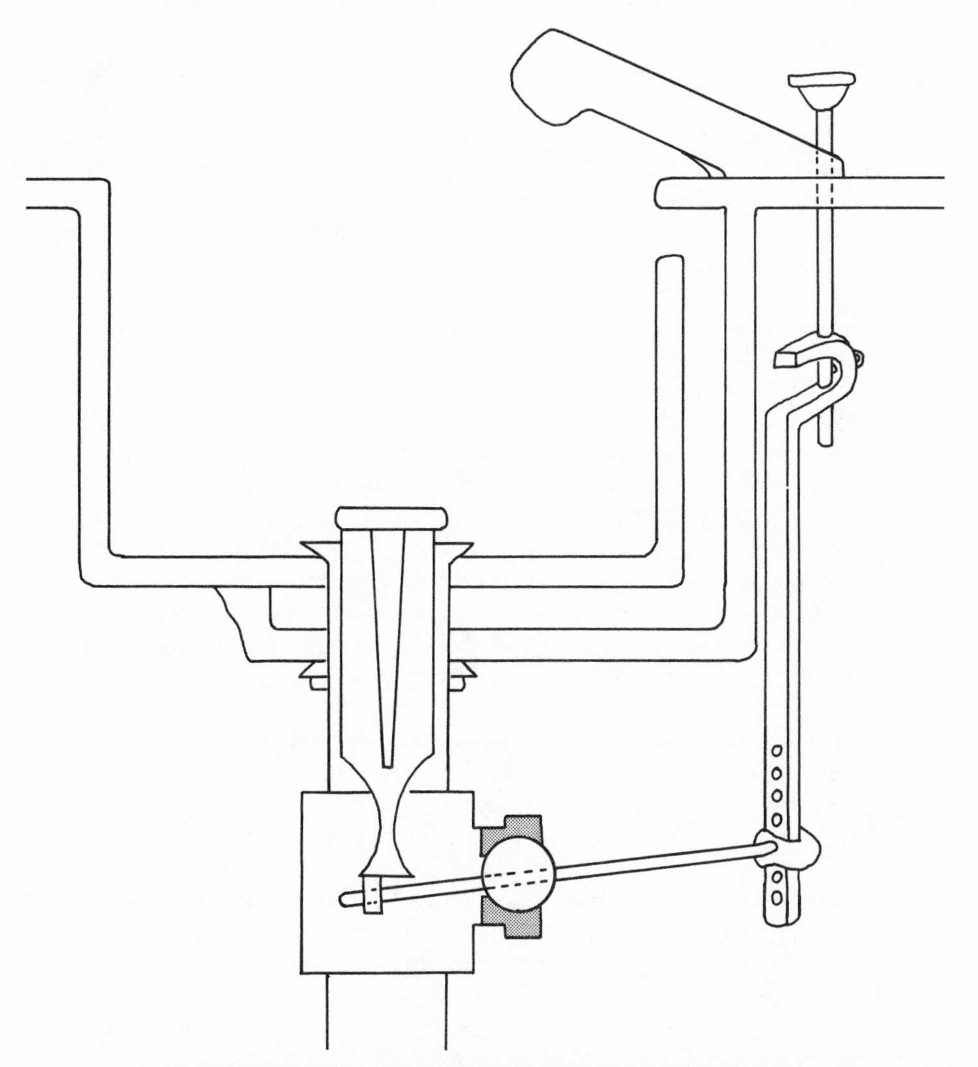

Cross-section of bathroom sink (lavatory), showing faucet and pop-up drain, with working parts, including a knobbed pull, attached to a perforated connecting bar, which has a "U"-formed end through which the pull passes, and which is attached to the pop-up mechanism in the tailpiece of the sink.

ing in the body which will allow the pop-up lever rod to come through. You may have to change the rod to one of the proper thickness or gauge for your new faucet, but that isn't always the case. The pop-up assembly lever can usually be adapted to any thickness rod, and nearly any system of connection to the new faucet.

If, on the other hand, you have been experiencing leaks from the seals of the pop-up lever in the tailpiece of your lavatory drain, it's best to replace the whole assembly. You can purchase a combination package, containing all the parts for your lavatory pop-up drain, as well as your new mixer faucet. *Read the manufacturer's instructions and look at the assembly diagrams.*

The faucet is installed in exactly the same manner as your kitchen centerset. If it has gaskets, you won't need a putty seal under the base. If it doesn't, make your putty rope and set the faucet.

SPECIAL COMMENT: The thicknesses of the ceramic or enamelware in different sink designs will vary. Lavatory tailpieces are designed to take that factor into consideration; they are adjustable.

Now, remove the old drain.

HELPFUL HINT: The types of tailpieces which incorporate the pop-up feature are made in two pieces. To remove one, loosen the hex nut beneath the bowl, and remove the entire lever assembly by unscrewing the lever packing-nut. Next, undo the compression nut to the p-trap. Then all that needs to be done (in most cases) is to turn the tailpiece, which should screw out of the p.o. plug or flange in the bottom of the sink. If it won't, then continue to unscrew the hex nut beneath the bowl, until you have the tailpiece nice and loose. Then push up on it and get a bite on the flange with your pump pliers. Turn the tailpiece out while holding the flange fast with your pliers.

Your new drain assembly goes back in reverse order; make sure that you putty-in the p.o. plug flange in the prescribed manner.

CAUTION: *When you are making up the new drain, be sure that you don't screw the tailpiece so high into the p.o. plug that the distance is too short for the pop-up to close all the way, when the lever is activated later. This is an apprentice's goof!*

When I troubleshoot leaking pop-up drains in new installations, that goof is the problem in almost 100 percent of the cases, and it only happens with the less expensive types of pop-up mechanisms.

The ideal makeup of this assembly is to screw the tailpiece up just enough so that its "drain slots"—which accept the water from the lavatory's overflow channel—cast in the ceramic or cast iron, are in the middle of the lavatory's channel, and *not* too high in the p.o. plug, which you can tell simply by looking down into the drain itself.

For sure, if your pop-up doesn't go down enough to seal when you want to stopper the sink, it is either because you haven't seated the pop-up properly on the lever in the tailpiece (check to make sure the slots or hook on the bottom of the pop-up stopper are properly set on the lever), or because the distance from the p.o. plug in the lavatory is too close to the lever pivot in the tailpiece. That can only happen if you've screwed the tailpiece too far up into the p.o. plug flange.

Now, if your faucets are properly connected, hot to the left and cold to the right, and your drains are effectively sealed, with the pop-up's mechanical linkages set at the required distances, the lavatory is effectively in service, and you are to be congratulated.

REPLACING YOUR TOILET TANK'S INNARDS

If your toilet spews water out the top of the tank when you lift the lid for inspection, shut off the angle stop and replace the defective float valve.

I don't think that there is any more exasperating leak than the one caused by arthritic, cancerous, or senile toilets. Sometimes they bug you in a way that is almost human. They clunk, leak, and hiss while you're asleep, and when you get up and fiddle with the pull-chain or handle, they clam up until you are comfortably snuggled in for a good night's sleep, at which time they inevitably resume their diabolical orchestrations.

At other times, you just can't get them to stop their continuous flushing action, without almost ripping the handle out in attempts to get a balky flapper or flush valve to close and stop Niagara Falls.

Add to those infamous goings-on the "hard flusher," the "quick closer," and the "tank dripper," and you have the score of a full symphony of irritating, money-wasting, but very common toilet afflictions.

Whenever you have just one of these ailments with which to contend, it is proper to go to the offending part and replace it. Don't even try to repair the internal tank parts of a toilet. They aren't meant to be repaired. They're designed to be replaced.

That means that, when the brass-wire link between the flushing handle and the rubber flush valve breaks or corrodes into ineffectiveness, you have absolutely no choice but to replace it *in kind.* The person who tries to bend a bit of wire coat hanger into a link is just asking for trouble. First of all, the coat hanger is steel, and the rest of the parts in the tank are brass and copper. Steel, submerged in water with brass and copper, creates a perfect battery, especially if you're into using those noxious and often ineffective tank "cleaning" compounds or solutions. You have created the perfect environment for the most degenerating of plumbing diseases, electrolysis.

The rule of thumb is that the farther apart the "index" numbers of the metals are, the faster an electrical current will attack those metals. Steel and copper are far enough apart on the index so that, given a water environment, the steel wire will probably corrode to uselessness in less than a month. Oh, it will seem to be sound in appearance, but the portion of the wire in contact with a brass or copper part will have developed "sticky" oxides which will inhibit free movement between the two metal surfaces. Even more insidious is the fact that the metal *lower* on the index is usually the one attacked. Steel is lower and is the metal which will deteriorate most rapidly. The brass and copper will also suffer, since, in electrolytic situations, the entire system is bathed in the debilitating electron flow.

This is the very same electrolysis that attacks galvanized steel pipe, especially when it is coupled to copper tubing without electrolytic protection.

Any iron or steel introduced into a toilet tank is going to be chewed to bits, and, in the process, will cause accelerated wear on the brass and copper parts. That is as certain as death and taxes.

When you hear the water hissing and "playing music" in your toilet, and you can't get rid of the tune, even with repeated jiggling of the flush handle, you are going to have to replace the "ballcock" assembly.

The ballcock is exactly what its name implies, a cock (another name for a valve) operated by a ball or float, so please don't rate this book "X" and hide it from junior's prying eyes, just yet.

If your ballcock is relatively new, then there is a simple adjustment to make that should return it to normal operation. CAUTION: *The first observation to be made is whether or not the tank water is spilling into the overflow pipe.*

The overflow pipe has many guises. Most often it is a brass standpipe attached to the casting in which the flush-valve ball seats to stop the flow of tank water into the bowl. Other overflow pipes are cast into the ceramic of the tank itself. Still others are plastic tubes formed with other plastic assemblies within the tank. One thing is

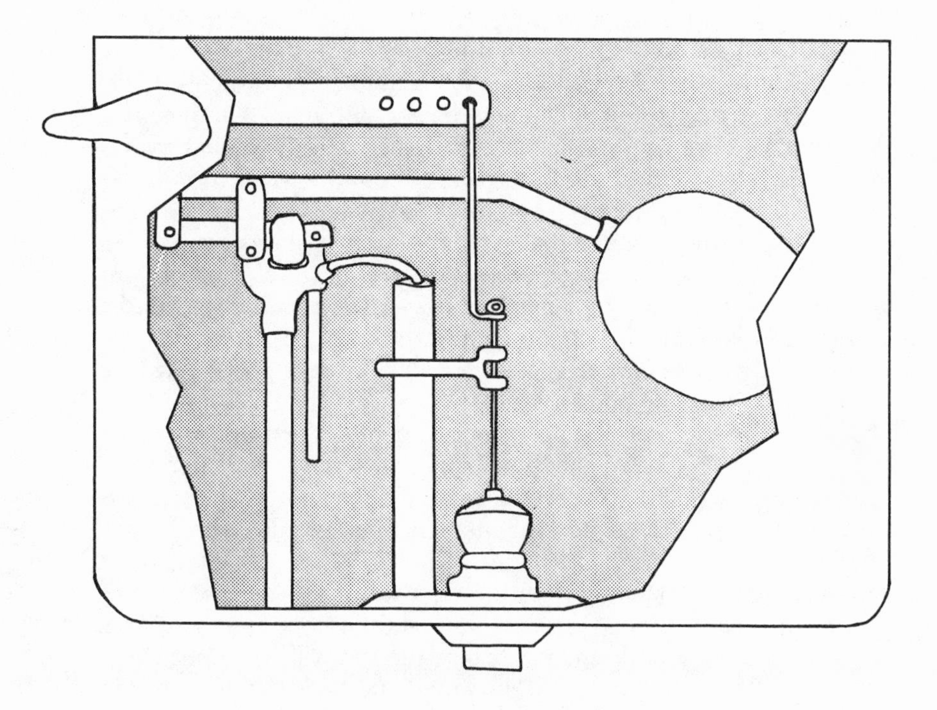

Cut-away of toilet (water closet) tank, showing parts of a conventional ballcock and flush valve.

sure, it's a tube into which the tank water flows when it has reached beyond its normal level for proper operation of the water closet.

If the water is spilling into that standpipe, the ballcock is either shot to hell, or completely out of adjustment.

There are three types of basic ballcock adjustments. The first is to make sure that the float rod is firmly connected both to its float and to the operating arm of the valve.

HELPFUL HINT: As you are surveying the interior of your water-closet tank, take notice of all water movement in and out of the tank, while all systems are as quiet as they can be, considering circumstances. The eddy and flow of incoming and overflowing water will give you quick answers to problems.

If the rod is secure to the float and the ballcock, and you hear water coming in at a slow rate, while small amounts of water are dribbling down the overflow tube, the float is set too high. If your ballcock is made of plastic, quite often there are small brass screws at the top of the valve that allow for restricted adjustment of the float to valve angle, or ratio. Failing that, flush the toilet and make a slight bend downward in the rod, while the water is still low in the tank. Then let the float fall to be picked up as the water rises in the tank. The ballcock should shut off *before* the water reaches the rim of the overflow tube. If it doesn't, repeat the last procedure.

CAUTION: *Having to make too great a bend in the float rod before the water stops hissing through the valve ballcock indicates a fundamental defect in the valve, and can only be solved by replacement.*

Some of the older, more expensive ballcocks have plugs and seats that can be removed and replaced, but frankly, the parts are rather hard to come by. Your best bet is replacement of the unit.

There is always the remote possibility that some particle or even a relatively large object has become caught between the plug and the seat of the valve. The third type of adjustment usually "cures" that problem.

Take the float rod loose from the valve arm. In some cases it simply means that you must unscrew the rod. In others, there could be a brass set screw to be undone before the rod will come out of its socket. *After it is out, turn off the water at the angle stop beneath the tank,* and examine the top of the valve. On some models, the top section can be unscrewed. Some of the plastic types are secured with a series of chrome-brass self-threading screws, with Phillips heads. Undo them and lift the top of the assembly off, gently. Make sure you keep all gaskets and parts.

The interior mechanism is simple beyond belief, consisting of a hole and a flap or plug to seal it shut and stop the flow of water. On the plastic types, the hole is closed by eccentric action on the back of the pivoting plastic flap. On the brass type, there is a simple linkage which lifts the plug and pushes it back into its seat as the float rises. If there are obstructions in the port or between the closure and its set, simply remove them manually, reassemble the valve and open the angle valve. Quickly reattach the float rod onto the valve's operating arm, and, as the tank fills, work the float gently up and down, which should cause the closure to reseat itself in its port properly. As you gently move the rod up and down, the valve should open and shut completely, stopping the flow of water. If it doesn't, replace the unit.

The flush valve has two basic configurations these days. The most common is the so-called "flapper," which consists of a hinged flap of neoprene or other rubbery material attached to a bulbous or stopper-like closure about 3½" in diameter. Some of these stoppers are a bit more sophisticated, having "water-delay chambers" and other exotic attachments, designed to allow the closures to remain open for a predetermined period of time before they drop into place and stop the flow to the bowl. Basically they're just flaps of rubber which fall onto a seat after the flush is completed, to keep the water in the tank from running into the bowl continuously.

These types are often attached to the overflow tube, which is equipped with pins or pivots to receive the holes molded into the flap assembly. Other types have hinge pins molded onto the seating ring or assembly. They are fundamentally quite simple.

The older flush valves consist of a flush ball, round on top and conical lower down. Some of them are open at the bottom, and other patented types terminate in a small, closed plastic cylinder. Into the top of this type a threaded brass rod screws, after it has been threaded through a brass bracket affixed to the brass overflow tube. The top end of the rod consists of an eye or ring. That ring is designed to act as a stop for the flushing link, which I've mentioned before in my sermon on electrolysis.

Nowadays that whole antiquated mechanism can be plastic as well, but it isn't likely. Most of the cello-packed replacements I've seen in building-materials stores have been the flapper type.

The only part which is apt to fail on the flapper-type is the rubber flapper itself. SPECIAL COMMENT: There is one type of flapper which is made out of rigid plastic material. It usually has a "water-delay chamber" and a soft rubber disc. Treat this type the same way as you do the flexible type.

The flapper type often cracks at the hinge holes and will not seat properly. It is free to wander when the first side breaks away from its hinge pin. When both sides part, there is complete failure.

In normal circumstances, all that must be done is to buy a replacement flapper, which usually comes with its own chain, and install it.

With the old-fashioned vertical rod-and-ball type, there are three failure points: (1) The ball itself can become worn at the base. It's easy to diagnose that problem. The rubber will be thin and very spongy. Often, it will be cracked and, when you run your fingers over the inside and bottom surfaces, they will come away stained with decomposing rubber. It's a simple matter to replace the ball. Just unscrew it from its rod, and screw on the new one. (2) If the rod is pitted or bent, replace it. That same suggestion applies to the (3) link wire between the valve-ball rod and the flushing-lever arm.

In stores that display rack cello-packs, the rod and link wire are usually packaged together.

BALLCOCK REPLACEMENT

Replacing the ballcock is a fairly easy exercise.

Replacing the complete flush-valve is a real chore.

HELPFUL HINT: There is a product on the market now that replaces the ballcock. It's called the *Fill Pro fill valve.* This ingenious, patented unit meters water into the water-closet tank without the need for a float. Needless to say, it installs cleanly and efficiently, reducing the clutter in the tank. CAUTION: *Because of code restrictions I've had reservations about using them. The use of the Fill Pro valve violates Section 1003(b) of the* UPC *which prohibits water-closet-tank cross-connections between the incoming fresh-water supply and the stored tank water.*

There is another type of toilet-tank fill valve on the market which does meet the code, called the *Fluidmaster* which is distinguished by a doughnut-type float which rides up and down the shaft of the valve assembly and is, therefore, relatively floatless when compared to a conventional ballcock. It also reduces the clutter in the toilet tank.

Fill Pro Fill Valve and Flap-Type Flush Valve

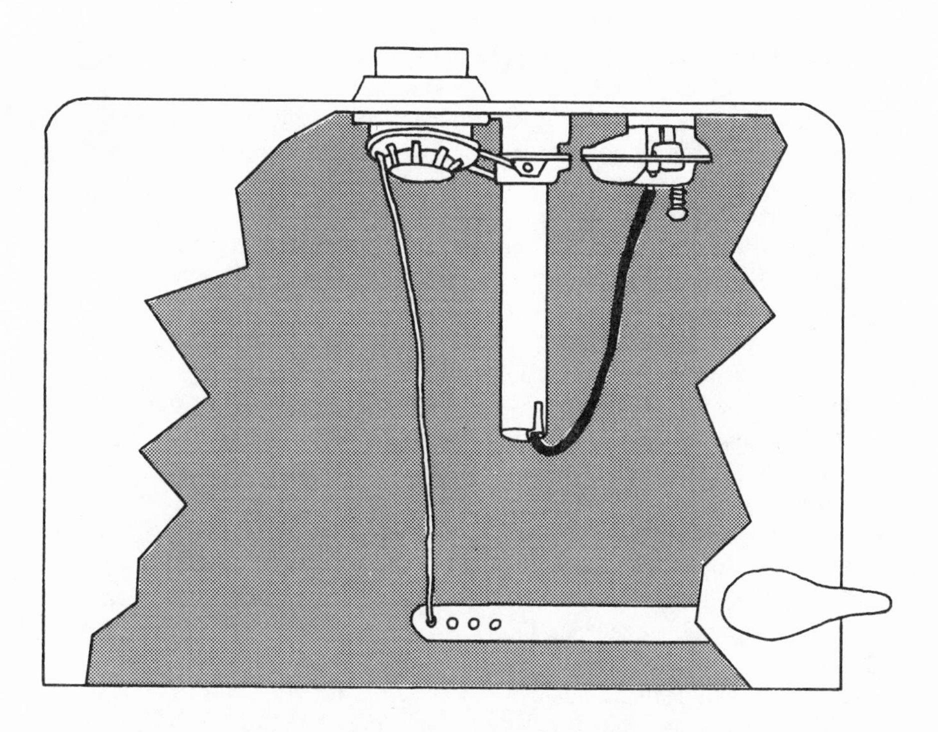

Cross-section of toilet tank showing Fill Pro tank filler and flap-type flush valve in place.

SPECIAL COMMENT: Clutter in the tank can be a distinct problem, especially when people insert weighted bottles or other objects to reduce the water volume. I have objection to those things and to others as well. A real culprit is the so-called sanitizer which is placed into the tank water in order to deodorize and "sanitize." I've been frightened by blue water turning to green before my very eyes on more than one "high" occasion.

Be that as it may, from all reports they're marginally effective at best, but people tend not to throw the containers away when they're finished. They just keep adding fresh units. Try to resist that habit.

I've seen gadgets come and go dozens of times over the last twenty years, and I've only recommended those new products which I've personally tested and found to perform as advertised.

I didn't buy the Fill Pro to replace standard ballcocks for my regular trade, in the beginning. Then, I began to get calls for them. Some people had bought and installed them themselves, and were recommending them to my steady customers, e.g. commercial building and apartment house owners.

Now, I install them, upon request, in all of the toilet tanks that I rebuild. I've installed them in apartment complexes over the past several years and I've never had a callback.

If your local dealer doesn't stock Fill Pro, you can probably obtain one directly from the manufacturer, J.H. Industries, Inc., 980 Rancheros Drive, San Marcos, CA 92069. If this sounds like a "commercial," forgive me. I normally don't push products, unless I am convinced that my readers will derive some real benefit from my suggestions. This is one such time.

To replace the ballcock or fill valve, simply turn off the water at the angle stop. Flush the toilet and then sop all of the tank water out, so that when you pull the ballcock, water won't run out the tank bottom onto your nice clean floor.

CAUTION: *If you have one of the super-deluxe, very-low-tank Case, Kohler or American Standard water closets, you will probably not be able to use off-the-rack ballcock assemblies, the Fluidmaster or Fill Pro. These types of toilets have very sophisticated ballcock assemblies, which can usually be rebuilt by knowledgeable plumbers.* Some of them have rebuild-kits. For those that do not have readily available replacement parts or kits, there is nothing to do but replace entire ballcock assemblies.

Undo the water-supply compression nut with an end wrench, crescent wrench, or your basin wrench. Then unscrew the retaining nut underneath the tank, and simply lift the ballcock out. You install either the ballcock or fill valve, in the reverse order.

FLUSH-VALVE REPLACEMENT

Before you get into the flush-valve replacement project, again be sure the tank is empty of water.

To replace the old-fashioned flush-valve-overflow tube assembly with the more modern flapper-type flush valve, you must first unbolt the tank from the water closet. If your tank is wall-hung, rather than close-coupled, your replacement is less tedious.

In that case, just unscrew the chrome flush elbow at the tank end, and then unscrew the retaining nut. Lift out the old flush valve from the inside of the tank, and you are ready to install your new flush-valve assembly. CAUTION: *With all installations in the toilet tank, you must make sure that the beveled gasket supplied with the part to be installed is in place inside the tank, or you will have leaks.*

Close-coupled toilets require the removal of the tank before the flush-valve assembly can be replaced. In most cases, you must unbolt three bolts underneath the tank, where it rests on the bowl flange (an extension of the toilet-bowl rim), to the rear. CAUTION: *Be very careful with wrenches in those places. The toilet is made of china and will crack easily.* FURTHER CAUTION: *As mentioned previously, before you work on a water-closet tank, be sure that you have taken the tank cover to a safe place. It is damn near irreplaceable, and should be handled with much "TLC."*

SPECIAL COMMENT: The tank-mounting bolts are usually brass and have large, flat, slotted heads that compress neoprene or rubber washers against the tank bottom. If those rubber washers are badly deformed, deteriorated or otherwise unserviceable, get new ones before you replace the tank on the bowl. If you haven't time to get down to the store, then when you replace the tank, set those washers in a bed of silicone, and allow the tank to stand without water for at least an hour before you refill it.

At this point, however, you've removed the tank. Now remove the beveled gasket which seals the tank spud to the bowl, and unscrew the large hex nut from the spud, which is actually a part of the flush valve. Insert your new flush valve, tighten up the new hex nut, replace the beveled gasket, give it a thin coating of silicone caulk, reseat it on the bowl and bolt it up.

Now, go through the usual exercises of reattaching the water supply, replacing the lid, and turning on the water (unless you have siliconed the bolt washers, which calls for delay).

Flush the toilet and check the water height. Make the customary adjustments to the float on your conventional ballcock, or on the water-level adjustment of your new Fill Pro.

The job is done.

LEAKS IN GALVANIZED-STEEL WATER PIPE

Steel pipe wears out much sooner than copper tube. That is because steel and iron oxidize rapidly in the presence of water. Most of your leaks are going to occur at couplings and other fittings, or at places where the steel pipe joins copper tubing.

The reasons are many, but the most probable are the ravages of age, poor threading or joining techniques during installation, or electrolysis (galvanic corrosion).

I have discussed electrolysis (galvanic corrosion) earlier in this section, and I will now cover construction techniques that can accelerate the deterioration of steel pipe.

The greatest problems, by far, up until now, have occurred in foreign-made pipe, especially pipe made in Far Eastern countries other than Japan, whose products are pretty much on a par with our domestic material.

Much imported pipe is too hard and brittle, making hand-threading very difficult. The dies don't want to grab to start, and once you've gotten them to "take," it requires two men and a mule to turn the handle on the threader, even on ½" pipe. Unless the pipe end is kept bathed in threading oil, continuously, while it is being threaded, the threads will break and slough out with the cuttings, leaving jagged, flattened, broken and imprecise threads. In order to get a good seal on these kinds of threads, it's necessary to take the fittings up to the limit of their taper. SPECIAL COMMENT: All steel pipe threads are taper threads, which means that they taper up to the full size of the pipe exterior, to create a wedge-tight fit when the fitting is taken up with a wrench.

This brittle foreign pipe threads so poorly that, in most cases, I won't buy it, if I can get American pipe. My next choice is Japanese; then Belgian or French, and from that point, most European and Commonwealth products. Korean pipe and products from nearby countries are the most difficult to work and thread.

Naturally, even when high-quality American pipe is threaded, the pipe section is thinned out in the thread valleys, and the galvanizing

Cutting Steel Pipe

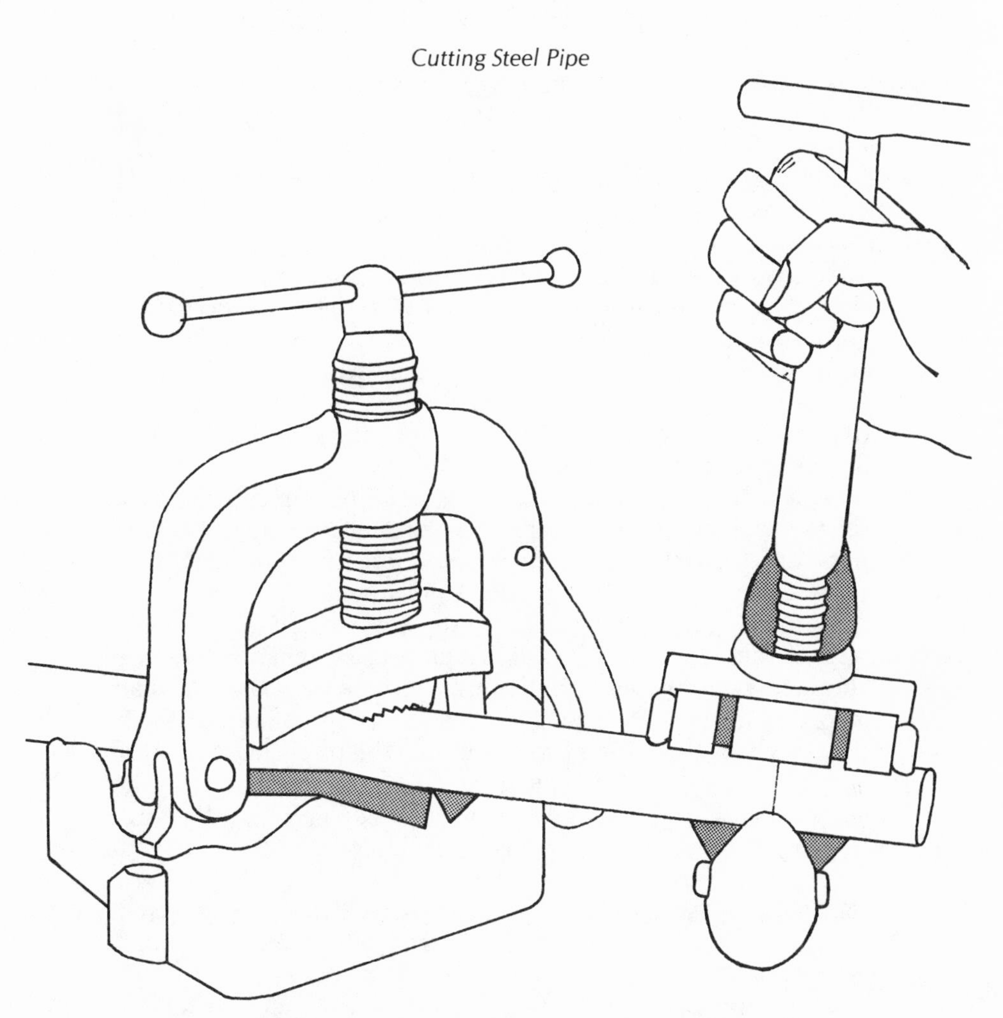

Steel pipe in pipe vise, being cut with a tee-handled pipe cutter.

which protects the rest of the pipe is removed by the threader. The inside threads of couplings and fittings are also ungalvanized for all intents and purposes, and what vestiges of zinc may remain in the threads are soon cut out when they mesh as the joints are made. The use of a high-quality pipe-joint compound, which also forms a protective bond over both the male and female threads, is your only insurance against almost immediate oxidation at the joints. Oxidation can speed up the process of weakening, from the day the system goes into service.

The lesson to be learned is this: If you use galvanized steel pipe in water services, make sure it is of the highest quality (American), that your threading machine dies are sharp and true, and that you join your pipe with the best pipe-joint compound you can buy, to minimize initial leaks and protect the threads.

To repair leaks in steel pipe at the joints, the only thing to do is take the pipe apart from the nearest union to the place of leaking, and replace the section of pipe going into the fitting. The fitting very seldom is the source of the leak. Fittings are usually made of galvanized malleable iron, which resists deterioration far better than the steel pipe. SPECIAL COMMENT: A pipe union is a special fitting that is used to join the ends of two pipes that have been put together from opposite ends, and that meet at a common location.

If there is no union within twenty feet of the leak, the most acceptable method of remedying it is to cut the pipe four to six inches from the leak side of the fitting on the offending pipe, turn out the bad section, insert a nipple of the proper size, and screw on a union. Thread the other side of the cut pipe in place with a small hand threader, attach the other side of the union it it, secure your union, and that's that.

HELPFUL HINT: I have not recommended that you purchase a pipe-threading stock-and-die set, pipe stand (pipe vise on three folding legs), or large pipe cutter in my tool list, because your use for these tools will be so limited that the investment is not justified. If you eventually replumb in copper water pipe, your need for such equipment will be nil. I therefore recommend that you rent these items from a contractor's rental, or equipment-rental company. They are listed under *Rental Service Stores & Yards* in the yellow pages of your telephone book.

If you find it difficult to thread the pipe in place, because of

Threading Steel Pipe

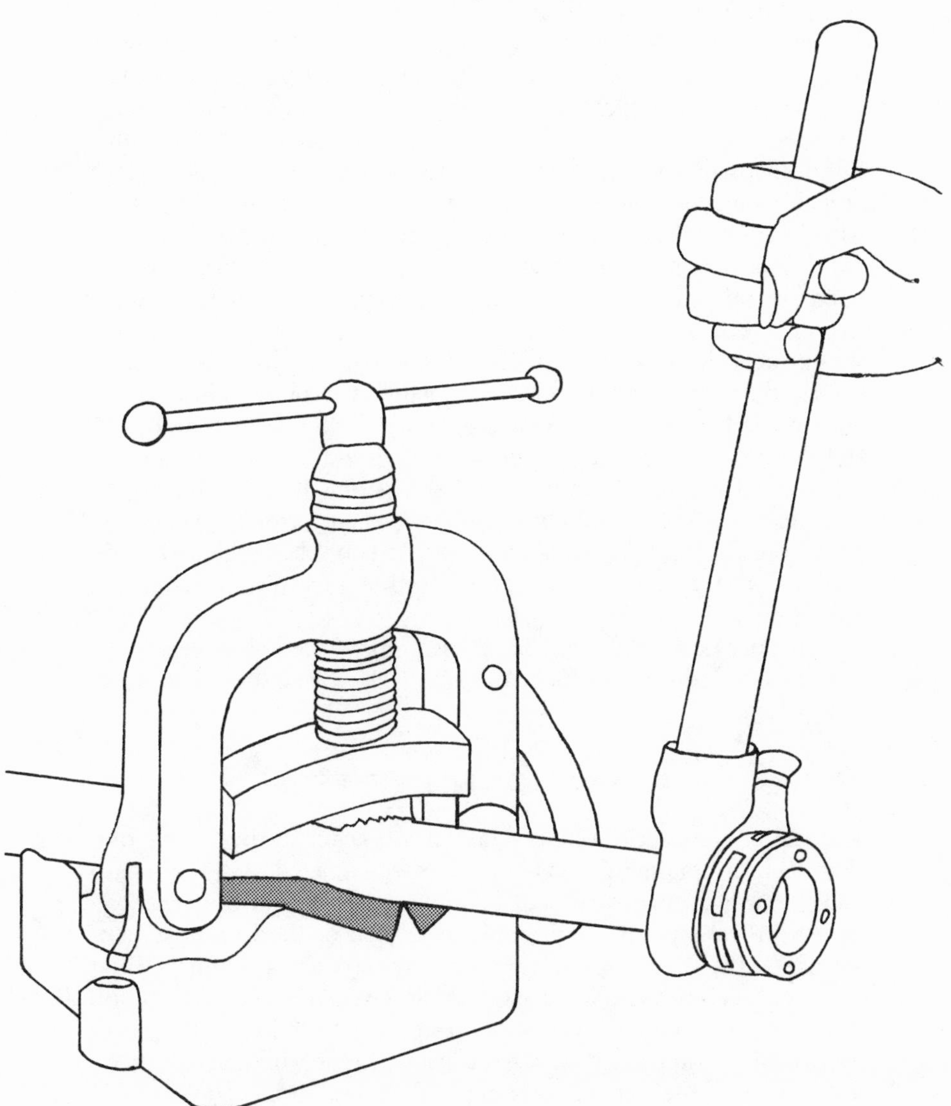

Steel pipe being threaded in a pipe vise. The threader is straight-handled, with a perforated and slotted cylindrical tip.

restricted working space, unscrew the pipe and do the threading outside, returning to complete the repair.

Leaks in the pipe itself are an indication that the walls of the pipe are about to go, not only on that length, but elsewhere as well. It is a sure sign that a complete replumbing of your water system will have to be done. These kinds of leaks usually occur as pinhole streams. For immediate, temporary repair go to your supply store and buy a "repair clamp" of the proper size. In most cases you will be dealing with ¾" and ½" pipe. In rare cases, your house main will be 1" or 1¼". Whatever size pipe is leaking, that is the size clamp you will need to stop it. Tell the clerk that you need a "pipe repair clamp for half-inch iron pipe," or whatever size is appropriate. The clamp should be a hinged sleeve with bolt-up flanges. This clamp compresses on a rectangular rubber patch which you place right over the area of the leak. Bolt up the clamp, and you should be able to control the leak. As soon as possible, replace the entire length of pipe. It's okay to wait until the next leak breaks out, if you wish, but be sure that you keep a watchful eye on the pipe from the day you install the repair clamp. Chances are that you'll be looking at more leaks within the month. The longest leak-free period you can reasonably anticipate is about one year.

If the leak occurs in the main, in the ground, the recommendation is that you read Section Five, and begin to replace the entire main right away.

CAST-IRON PIPE LEAKS

Leaks in cast-iron soil pipe almost never occur in the middle of a pipe section, unless that section has been damaged by a very sharp blow from a heavy instrument of some sort. Of course, cracks can appear in the pipe, if the ground has shifted or there is movement for which the pipe bed had not been properly prepared in the first place. A good example of that is a shallow initial burial, based upon the premise that there will be no traffic over the area, other than foot traffic or an occasional wheelbarrow. Later, you install a driveway over it. If the ground is spongy, and you don't lay a proper base around the pipe where it crosses your driveway, or better yet, bury the pipe deeper in the ground if possible, it will crack eventually, just from the pressure of the new traffic.

A crack lets in roots, and roots fill sewers, as I mentioned before.

A leak at a joint, if it is to be repaired in a truly professional manner, would require yarning and caulking irons, a propane- or kerosene-fired lead pot, some tarred-rope oakum, a pig or two of lead, perhaps a dam (the tinker's dam of legend), and a lot more knowledge than is possesed by most young plumbers who have taken their journeyman's ticket in the past five years. Some union locals don't even teach apprentices how to lead-caulk bell-and-spigot soil pipe nowadays. It is still done in some areas, but they are becoming few and far between since the acceptance of no-hub soil pipe in the code.

HELPFUL HINT: In plumbing supply stores there are two wonderful products available that are usually not available in do-it-yourself stores.

1. Two-compartment epoxy plumber's putty, which usually comes in a one-pound package consisting of two sticks of putty-like dough, one white and one colored. To use this material you simply cut off two equal pieces from both sticks. I usually just lay them down side-by-side and cut them with a sharp knife like two pieces of hard butter. CAUTION: *Be sure to clean your knife blade well, with a solvent-impregnated rag. The two substances could have made a light mix as you drew your knife through them both at the same time. Once that stuff hardens, forget it. It's stronger than concrete and many times as difficult to remove from anything. Have patience mixing the epoxy putty. Make sure that you don't just have fine lines of color running through the white dough. The ball of material which you knead must be colored completely through, evenly.* The best thing to do is leave the box of putty in your kitchen or where it is warm for a day before you use it. CAUTION: *Do not heat it over a flame or electric burner.* Let it warm up to room temperature in the environment. It can be worked in cool condition, but it will be hard on your hands and wrist muscles, until the two materials begin to work on each other, at which time they will generate warmth internally, and will work quite easily.

2. Epoxy paste is usually made by the same manufacturer as makes the putty. This paste works with a spatula, and is very effective on hairline leaks.

CAUTION: *Wash your hands thoroughly after using epoxy products. If you are allergic to skin irritants, use gloves.*

To prepare the surface to be coated or packed, get an ordinary large-blade screwdriver, one that you don't care much about any more. Actually, any stout screwdriver-like blade will do. You'll need a hammer, to go with the blade tip, and you're set.

Go to the offending joint, and begin tapping the tip of your blade into the lead seal. Start 180° away from the leak, if you can see where it's leaking. You're beginning on the opposite side of the leaking or weeping section. Give the blade a whack, but not so hard that you warp the band of lead to the point where it starts to lift toward the edge of the bell, as a band or ribbon. If that happens, stop!

CAUTION: *Out in the country, where a journeyman plumber was scarce in the "old days," some pipe was installed with packed oakum and cement mortar. There are pipes which have been sealed with sulfur-mortar mixes of sorts. And there is lots of soil pipe just held together with packed oakum and hot tar pours, called bituminous packs. If your joints have anything but oakum and lead seals, the technique I am describing is useless.*

All you need for the others is the epoxy.

Back to the screwdriver and hammer. Let the blade lie across the span between the pipe and the inner wall of the bell, wherever possible, and keep going, tapping the lead in tight to make a seal between the pipe and the bell. Tap evenly, and set your screwdriver blade no further than ¼" ahead of the previous tap, as you move along. Go around both sides of the pipe *toward* the leaking section, ending up with the leak point itself, where you can concentrate your remaining blows. Quite often, just this procedure will seal the leak.

HELPFUL HINT: Before you apply the epoxy, run water into the sewer line from the nearest fixture to the leak. First run the cold for about three minutes, and then run the hot for an equal time. This should cause the line to contract and then expand, giving the joint a good workout.

Frankly, in most cases, you will not have stopped the leak.

Now, turn off the water, prepare your epoxy, and then roll it into a rope, as with ordinary plumbers' putty. Lay your epoxy rope into the bell of the leaking joint, and stuff it in with a small piece of wood, formed especially for the job. A small piece of thin boxwood, just wide enough to smooth the putty in the bell joint, is just fine. It is a good idea to double-check by inserting your small finger into the space and smoothing the whole band of epoxy against both the pipe and the bell wall. It is also a good idea to add layers of the epoxy until you've filled the bell entirely, so that you can smooth it flush to the bell rim, but it isn't really essential. The additional "lifts" of material just add insurance against future problems.

SPECIAL COMMENT: This putty is expensive, costing between $10 and $15 per pound at this writing, but the savings in terms of saved plumbers' costs would be phenomenal, even if you used the whole pound.

For the other types of joints mentioned, you can use the epoxy putty, without any prior tamping with your screwdriver, or, if the bell is filled to its rim with the original caulking material, paint the entire bell area, up the pipe about 1" and down the outside of the bell an equal distance, with the epoxy paste, in two or three coats. The alternative to that is to create a solid band of epoxy putty to cover the same areas. Either way you'll get a positive seal.

If epoxy is unavailable to you, the best seal I know is Grandma's old standby, canning paraffin. In a vertical pipe, where the joint forms a cup, just pour the cup full of paraffin, but before you do that, make sure that there are no wide-open cracks or slots for the wax to run down into the pipe. You can seal those with a bit of newspaper or heavy twine. The horizontal pipe, with its side-lying joint, is the real problem, but there is a good solution. You must create a dam with masking tape.

Get some 2" masking tape and stick one edge to the bell rim and the other to the pipe, all the way around the pipe, except leave a pour opening at the top, into which you'll pour your melted wax from any kind of pitcher-like vessel.

HELPFUL HINT: If you're working under the house, and it will take some finagling to get the hot wax from your stove to the point of use before it congeals, just set a small Sterno stove under the house and melt your wax there. A metal measuring cup, equipped with a pouring lip, is the best vessel to use. It's even worth the investment, if it will save a trip by the plumber.

The only ways I know to stop a leak in a cracked section of soil pipe are either to remove the broken section, which is a major job, or to support the cracked section against future damage, and create a collar out of one or two pounds of epoxy putty. That collar *must* be a complete band or circle around the pipe to give it the support and strength of integrity of material. It should be no less than 2" wide and ½" thick.

If you have room to work under the house, you want the practice and don't mind the aggravation, go to your rental shop and rent a "snapper-type" soil-pipe cutter. This particular tool resembles an ordinary pipe cutter in some ways, but essentially it consists of a stock or handle, a ratchet, and a chain containing twenty or more cutter wheels. The principle of this cutter is to surround the pipe with cutter wheels, which are then drawn down on the iron pipe surface by ratchet action. You only have to rock the cutter on the surface about an inch each way to create a groove. Then, you ratchet the handle until the pipe snaps under the wheels. When you remove the tool, you'll see a very nice, clean cut, with a bunch of sewage running out of the crack. CAUTION: *In order to avoid getting a mouthful of toilet water as you crack the pipe, you'd better have everyone in the house go to the toilet* before *you cut the pipe. Then, shut the angle stops to every fixture which empties into the line you're cutting.*

Thereon hangs a tale:

I was the newest journeyman in a large shop in San Francisco many years ago, so I had the weekend trouble-call duty, since our shop advertised 24-hour service.

About three o'clock, on a rainy winter morning, I received a call from the boss to go out and take care of a stopped-up sewer in Saint Francis Woods, a very posh district, largely populated by high-ranking politicians, old upper-middle-class San Francisco families, consular officials and other nabobs of "quality," in those days. I was just supposed to go to the house, assess the situation (another way of saying, put in an appearance so the shop could charge for an "overtime" call), and report back to the boss.

When I got there, it was pouring rain, every light in the three-story mansion was lit, and there were people in bathrobes running around, pulling their hair. The upstairs bathroom was under two inches of foul-smelling water, with human waste and paper floating out in the hallway. Daddy had tried to use a plunger on the stoppage, and the very expensive commercial-type flushometer valve had stuck in the open position, continuing the flood.

Someone with presence of mind had shut down the house water main, so there was no running water in the house.

Everyone descended upon me as I rang the doorbell. For a moment, I felt like the Archangel Gabriel, expected to blow his horn and put everything to rights.

I couldn't get out of it. I had to do something on the spot. Hell, I was twenty-six years old, and new on the job. What did I know? I called in and got orders to clear the stoppage and clean up the mess.

I tried everything. We didn't have Drain Kings in those days, so I went down the line of options from a closet auger, which didn't work, to taking the water closet off its ring and using a motor auger right down the stack as far as it would go, which was fifty feet, and then down to the cleanout for a shot through the main. The stoppage was in the 5″ main. It was a bitch.

Whatever was in that line was as solid as a rock, and it later turned out to be plaster-like hard-setting material, flushed into the line that morning by the gardener or handyman. I dug up the line, and cut out the plugged section, after I had warned everyone in the house not to flush any of the other toilets.

Someone, who shall remain nameless, didn't heed my plea, and while I was leaning down inspecting the open end of the line, I got a turd in the face, my first and last.

The irony of that was the fact that the offending toilet was the *only* tank-type water closet in the entire house. If anyone had tried to flush the flushometer-equipped water closets they wouldn't have worked, because the water was off.

To this day, I can still remember every small detail of that experience. *Beware when you work on sewers.*

So, you've cut your line and you're ready to make your repair. Now, cut out the section of the pipe which is leaking. CAUTION: *It is difficult and impractical to cut out less than a foot of pipe for the type of repair I'm about to describe.* FURTHER CAUTION: *If the pipe you're cutting is vertical, you must support the upper section, or it will tend to drop down when you remove the piece to be replaced. The best support I know is plumber's perforated metal tape, taken around the pipe and nailed to any available surface. This will place a "friction drag" on the pipe, which should hold it in place.*

The plumbing supply store should have no-hub cast-iron soil pipe in various sizes from 2″ to 6″. The shortest lengths available are 2 ft. and 4 ft. Most shops don't bother to stock the 2 ft. length, and 5 ft. and 10 ft. lengths are standard. Whatever size you have to buy for your 1 ft. repair section, buy it. You may need the rest later. At the same time, you'll have to get two no-hub couplings, consisting of a neoprene or other rubber-type material sleeve and a stainless-steel clamping oversleeve coupler, which is tightened with a ⁵⁄₁₆" nut-driver torque wrench.

Again, you'll have to use your "snapper" to cut the proper length of "patch" line.

HELPFUL HINTS: Make sure that you cut your old line so that you have about a ½" working gap. That means that you will have a 12½" space to receive your one-foot section. Now, put your neoprene sleeves on both ends of the new section and force back the portions which are not on the pipe itself. With practice and patience, you should be able to get the rubber to stretch enough to allow you to roll the sleeve over on itself so that you now have a smooth round bead on each end of the pipe. It is very difficult to get the sleeve to roll all the way back on itself to the seating band in the middle of the sleeve, so you allow that ½" to permit your new section to pass into the gap between your two old pipe sections, where you can then promptly roll both sleeve ends back to cover the old pipe ends. *Important: Before you insert the new section, be sure that you have already placed your stainless steel sleeve clamps on the pipe, so that when you've snapped the rubber sleeves in place, you can cover them with the cinch-up clamps, and make the joint.*

If you forget, no sweat. You can always take the clamp bolts loose, slip the open clamp sleeve over each connection, refasten the bolts, and take them up, completing the joint.

HELPFUL HINT: If you don't have access to a one-foot section of pipe near the leak, consider taking loose an entire assembly, and replacing it with no-hub fittings and sections. In such cases, you will have to "snap-cut" several sections, often of different sizes, but if you must, you must. Fittings of exactly the same radii and design as the bell-and-spigot fittings in your system are available from your plumbing supply house.

Don't be afraid. Take your time, and, if possible, have help. Iron fittings and pipe are heavy.

CAUTION: *Make sure all water is cut off to the lines involved, and all toilets are flushed, before you open sewer lines.*

BITUMINOUS AND CLAY PIPE LEAKS

Bituminous and clay pipe may be repaired in much the same fashion as cast-iron pipe. The same "snapper" with which you cut iron will work on clay as well. Failing that, get a masonry blade for your power saw. It will do a satisfactory job on clay pipe.

Bituminous pipe, which is used mostly in septic tank systems, looks very much like it is made from tar-impregnated paper layers, compressed to form a tube. Of course, that is because I have described the process of making it.

You can cut bituminous pipe with saws of almost any type. Joints are made with special sleeves.

There are clamping sleeves made for clay pipe, which very much resemble those used on iron pipe.

A quick repair for sewers in the ground is to dig around the repair location, and pour it solid with concrete, using the smallest available gravel as aggregates. Such a repair may eventually seep a bit, but it will usually hold to reasonable tolerances.

CAUTION: *Whenever you pour a leak repair, be damned sure that the pipe on both sides of the repair is resting on well-compacted, undisturbed earth, and that your pour is into cut and otherwise undisturbed earth. If those conditions do not exist, the weight of your repair may crack the pipe on both sides if there is earth movement or settling, as a result of the repair. Soft earth gives, which can be fatal to cast-iron, clay, and some types of bituminous pipe.*

PLASTIC- AND COPPER-PIPE LEAKS

I haven't dwelt upon repair to plastic and copper pipes and tubes because usually the only leaks that occur in these products are the result of mechanical damage rather than deterioration in service.

For breaks in or other inadvertent damage to copper pipe or tubing, use repair clamps, or read Section Five on the soldering of copper pipe and fittings.

For damage or leaks in plastic pipe, a simple section or fitting replacement with couplings and solvent cement are the usual remedies. Section Five describes plastic pipe assembly and joining techniques in detail.

Leaks in your rainwater and storm-drainage systems require pretty much the same techniques and remedy options discussed throughout this section. They are covered in depth in Section Four, which follows.

GAS-PIPELINE LEAKS

Gas leaks account for the fewest problems in steel pipe. That is because there is no water flow in gas lines, except condensation, which is not a great factor in long-term corrosion.

Leaks in gas lines are most apt to occur at fittings, in valves (called gascocks), but most occur in the connectors between the steel lines and appliances.

For a great many years, corrugated or accordian-pleated copper connectors have been used between gas pipes and stoves, heaters and furnaces.

If you *ever* move an appliance for purposes of cleaning behind it or replacing it, or for any other reason, check for leaks after you have returned or replaced the appliance in its final location.

CAUTION: *Do not use matches, cigarette lighters, or any other device that creates a spark or open flame to check for gas leaks.*

HELPFUL HINT: The best device I've ever discovered to use in gas-line leak checks is an ordinary squeeze bottle with a solution of 40 percent detergent and 60 percent water. A couple of ounces of the solution will do quite nicely.

Simply squirt small amounts of the solution at each end of a fitting, gascock, or connector. Wait for a minute or so and go back over all the joints you've treated. Leaks will show up as clumps of fine bubbles on the leaking area.

THE REASONS: The joints were not taken up tightly enough when they were first made, and the pipe-joint compound has dried and shrunk, creating the proper environment for a leak. The pipe or fitting may have been damaged by someone trying to force the pipe aside to make an entryway or passage. The pipe or fitting may have been whacked by a vehicle or piece of equipment. If buried in the ground, the soil may have been acting upon the surface of the pipe, breaking down the protective coating and attacking the steel or iron; this is especially true if the soil is very acidic or basic (alkaline). The joints may have been loosened when a line was taken apart to add a new fitting or valve, and then tightened up again, without new pipe-joint compound.

In flexible accordian-pleated connectors of the old type, just being flexed once too often works cracks into the tubing, which can leak. Be sure that, if your flexible connector is of the old type (bare uncoated copper), you replace it with a new style, approved plastic-coated connector.

SPECIAL COMMENT: The *UPC* permits connectors of up to 36″ in length on all appliances, except ranges. Your kitchen range may utilize a connector of no more than 72″ in length. They must be of the same nominal size as the *inlet connection* of the appliance.

You are not allowed to hide these connectors in the walls. You may not put them through a wall from the pipe on one side to the appliance on the other side of the wall.

CAUTION: *Make sure that you have a gascock mounted on the pipe before you attach the connector, so that if you need to change either the appliance or the connector, you can turn off the gas near the location of the work. Do not use rubber gas hoses in the house. They are for outside use only.*

Aluminum connectors may be used, according to the code, but I don't recommend them. Any kind of oxidant will attack aluminum, including sea air, bleach water, some types of mineral-rich potable water, plaster splatter, and a rather long list of common household chemicals and detergents.

As I've said before, outdoor appliances may be connected to your house gas service by means of an approved outdoor hose connector, to a limit of 15 ft. from the permanent gas line.

The only exceptions to this rule are laboratory burners and equipment that requires flexible gas connections. Rubber hoses may be used for them, but they cannot exceed 6 ft. in length, can't go between rooms, and must be in sight for their *entire* lengths.

The remedy for leaking joints, except where the leaks have been caused by obvious corrosion, is to take the line apart to the place where you can separate the leaking section (usually at a union), recoat all pipe threads with an adequate amount of good pipe-joint compound, and rejoin the pipe, taking it up snugly at each joint with a proper-size pipe wrench. For pipe up to 1″, an 18″ Stillson or Ridgid-pattern pipe wrench is quite adequate.

CAUTION: *When you've taken the pipe apart, check the fittings themselves to make sure there are no hairline cracks or thread defects.* It is rather rare, but I have seen defective fittings that have held gas for years and then have failed.

HELPFUL HINT: I usually replace every fitting and gascock on a reinstalled line.

Don't fiddle with leaking gascocks, controls, or gas-served devices of any kind. If they leak, replace them.

Beware of yellow, sooty flame. It can be dangerous. A properly adjusted range, furnace, stove or oven will have a flame that just touches the burner surface, rather than dancing on air a bit above it. It will be predominantly blue, almost completely smokeless, and will show just the hint of red and yellow on the tip of the flame. If you can get all the yellow out of it, and still have the base of the flame resting right on the burner openings, you have adjusted it perfectly.

A lot of hissing noise, with the flame base in the air above the burner is an indication of either too much gas pressure through the utility company's regulator or a poorly adjusted air-gas mix. You can regulate that yourself. Look at the place where the gas enters the burner assembly. Usually there is an adjustable air intake consisting of a round plate or a rectangular shutter, held onto the burner assembly by flat-head machine screws that can be loosened and tightened with an ordinary screwdriver.

After you've loosened the screws you should either be able to rotate the round plate or slide the rectangular shutter, to allow more or less air to enter the burner arm. If you let in too much air, the flame will dance above the burner surface, and there may be a bit more hissing. If you don't let in enough, the flame will be mostly yellow with flecks of red. A blue flame with a brighter core than outer edge, with the hint of yellow or red at the tip, is pretty much where you want it to be. When you've found the right place, tighten up the screws and let 'er rip.

Remember: After you've retightened your gas line, run the bubble check on every new joint you've made. In fact, if you can, leak-check the whole line, front to back, top to bottom.

WATER-HEATER LEAKS AND REPLACEMENT

The day you happen to notice a damp spot around or beneath your water heater should be a day you mark on your calendar, to remind you to check the heater more frequently from then on.

Wetness around the water heater can indicate the beginning of a major problem or it can be harmless condensation, dripping down from overhead pipes or from the outer tank surface of a heater that has just been returned to service after a long period of disuse.

Condensation on water-heater pipes is very common. It usually happens in copper plumbing, where the surface change from hot to cold is very rapid. Very cold water, coming into the heater on a warm day, will frost the pipe like the outside of a glass of ice-cold water. That is condensation. It can gather, drop to the floor, and reform to drop once again, in a continuous cycle. The only way to prevent it is to insulate the cold-water pipe a few feet before it enters the heater.

HELPFUL HINT: I'd like to suggest that you insulate *all* of your hot-water lines as well. You'll be surprised how much money you'll save on your energy bills. Don't worry about the pipe in the walls. Just insulate what you can see and get to easily. If you can manage 50 percent of the total hot-water piping in your house, there will be a direct saving of as much as 10 to 15 percent of the heat loss between your heater and the farthest fixture on the line. That can translate to a lot of bucks over the years.

If you determine that there is no condensation, or pipeline leaks near your heater, and if it has been in more-or-less continuous service for a reasonably long period of time, and you still have a damp area around it, chances are that your heater is beginning to go. There may be a hairline crack or pinhole leak in the tank.

The problem may go away for awhile, and you may think that the dampness may have been caused by any number of other things which can and do happen in garages, basements, kitchens, and furnace rooms, but you should remain on the alert.

Quite often, pinholes and cracks self-seal, or literally rust tight. The oxidation will act just like a patch for awhile, but when that temporary dam finally gives way, there will no longer be any question in your mind, because the floor will not just be damp, but wet, even to the point of minor flooding.

Sometime before that final event, however, the water will have run into the firebox and quenched the pilot flame. Your first real inkling that something is radically wrong may come when you try to take your morning shower and end up all covered with goose bumps and on the edge of a serious chilblain attack.

After you've gnashed your teeth and yelled loudly enough to serve notice upon all creation of your intent to murder the person who failed to pay last month's gas bill, your best diagnostic bet is a glance at the water heater.

Often, the problem is as simple as a blown-out pilot flame. The answer is obvious. But, before you try to light the pilot once again, check the firebox, and make sure that it isn't full of water or debris.

If the floor is wet, *you need a new heater!*

A plumber will charge you between $65 and $235 to install your new water heater. If he furnishes the heater, you may be looking at the better part of $500, labor and materials, before you're ready to take your next hot shower.

Before you do anything else, shut off the water valve, which is usually located on the cold-water line going into the heater. Some heaters have two valves, one on each line. That's nice but not essential.

After you have turned off the cold-water valve, go around to every fixture in your house which has a cold-and-hot-mixer faucet. Put a little sign on the hot-water handle which states plainly "no hot water today," or "out of order," or anything else you can think of that will be effective with your family. Then tell everyone in the house what has happened and caution them *not to turn on the hot water handle on mixer faucets.*

THE REASON: If the hot-water side of a mixer is open, when you turn on the cold water, a portion of the cold water will enter the hot-water lines, especially if you have a nearly stopped-up aerator on the faucet spout, as many people do. If, for some reason, a faucet spout becomes clogged, it could back cold water up all the way to the heater and start it leaking again. Worse than that, if you have removed the leaking heater and have not yet replaced it with the new one, it is almost certain that water will back up and out of the hot-water pipe, with the slightest backpressure, from any place on the same level as the heater. It doesn't even take backpressure if the fixture is above the heater. A certain amount of water from the open hot tap of a mixer will blow out the hot side of the missing water heater's line by simple gravity.

That's why I always place two valves on every water heater I install, one on the hot side and one on the cold side. It saves mess and a lot of aggravation later on, when the heater or pipes need replacement or repair.

At the same time as you shut off the water to the heater, please close the gascock.

Water-heater replacement can be either a piece of cake or the most infuriating chore you have ever attempted.

If your heater is located in a nice open space in a ground-floor garage, you've got it made.

If the offending heater is in a small closet on the third floor of a turn-of-the-century gingerbread Victorian, with a narrow staircase covered in frayed, sliding old carpet, it will be futile to try to do the job without help, unless you have a great deal of rigging experience, as well as courage "above and beyond the call."

HELPFUL HINT: If possible, move the heater to a new, more accessible location. The information here, combined with construction techniques discussed in Section Five, should assist you in that project.

Unions should join the old heater to its piping. Of course they must be undone to remove it. After you've gotten them loose, disconnect the gas line from the heater controls. CAUTION: *Be sure that the gas is shut off completely, by turning the gascock handle to lie across the gas line. If your gascock doesn't have a lever handle, make sure that it is fully closed according to whatever indicator is usual for that valve type, before you go any further.*

HELPFUL HINTS: Often the unions on the water lines to and from the heater are "frozen." If your largest wrenches fail to budge the unions, use heat, as I suggested before, in the section on kitchen-sink centerset replacement. The object is to heat the big nut holding the two segments of the union together. When you have it so hot that it's smoking, give it a solid whack or two with your hammer, and get a wrench on it as quickly as you can. Give it a good hard tug. If that doesn't work, try the procedure again. If, after three or four attempts, the union still won't "break," you're going to need some help, and a "cheater."

A "cheater" is a pipe which fits over the handle of your largest wrench. There is usually some scrap around most older homes. If a thorough search doesn't turn up a length of old pipe that will work, then wire an iron rod or piece of wood to your wrench, with good solid windings of stone wire or bailing wire, the farmer's "friend."

Get a friend to help you by bracing the water heater as you apply the pressure of your wrench to the union.

One last heating, tapping, and wrenching with the "cheater" should do the trick.

If it doesn't, cut the pipe above the union, and replace it with a new piece when you connect the new heater. An ordinary hacksaw will do a fine job of cutting steel pipe. Your copper pipe cutter will handle any cuts on copper risers.

The gas line should come loose without any difficulty.

With the water heater disconnected, attach a hose to the drain bibb at the bottom of the tank, and empty it completely, before you try to move the heater. Water weighs 8.33 lbs. per gallon. That means that a "little" 30-gallon water heater contains 250 lbs. of water. Added to the weight of the heater, that can be a regular gusset-buster.

HELPFUL HINTS: Normally, it is difficult to move a water heater without assistance, until you are instructed in the "walking technique." The easiest way to walk a heater out of its place is to use the two ¾" hot and cold nipples that should still be in place, as handles. If you've removed them, put some 6"-long nipples in place, and grasp them with both hands, tilting the now-empty heater back against your shoulder, just enough to get one or two of the heater feet off the floor. Now, try to rotate the tank either to the left or right. Notice that in rotating the unit, you have also moved it toward you slightly. Let it down on all of its feet once more and then repeat the tilt-and-roll movement to the other side. And so, you'll be, in effect, walking the tank on its feet, moving it toward you a bit each time you tilt-and-roll.

If the space is too narrow to walk the tank, then you'll have to tilt it, holding onto your nipple-"handles," and let it come down to where your arms are straight down. Check and see if the controls and the drain bibb at the bottom of the tank are clear. If they aren't, lift up slightly. Then, give a good tug and it should come out, sliding on its legs.

Of course, if the heater is in a closet off your wall-to-wall-carpeted main hallway, dragging the tank or even walking it, is out of the question. You're going to need assistance from a strong person, who is slender enough to get behind the tank, if possible. If that is impossible, then, the strongest person picks up from the bottom while the other guides the top. The object is to get the tank down on the floor, without soiling or damaging the surface. That means having a nice bed of newspapers or a plastic garbage bag opened up beneath a pile of rags, or a similar protective combination.

CAUTION: *If you have been unable to remove* all *of the water from the tank, don't let it become horizontal, or the rusty, sludgy bottom water will probably come right out of the top spuds and soil everything in sight. The only way to protect yourself from this is to screw the nipples in solidly with pipe dope (joint compound) and then cap them with ¾" caps, tightened into place with more compound.* Also, *make sure the bottom drain bibb is shut off all the way, and securely wrapped in rags and plastic, making a moisture barrier. Don't allow the wrapped bibb to touch the floor or you'll peel away your protection.*

HELPFUL HINT: Believe it or not, a sturdy children's wagon is often a very satisfactory "dolly" for moving heaters of up to forty-gallons capacity, from one place to another. When I say "sturdy," I mean it must carry weight of up to 150 lbs. One of those inexpensive "flyer" models, with little metal and bracing, just won't do. It's best to try the wagon out by sitting in it yourself and seeing whether or not the axle brackets and other running gear are able to bear the weight and move. In most cases, if the wagon will handle 125 lbs. of person, it can take the weight of most thirty- and forty-gallon water heaters. The problem will be getting it in place, and then keeping it aboard for the trip. I suggest that, if you can't tip the heater down onto the length of the wagon, so that the tank lies over its long bed, you probably won't be able to make it work. If you can get it to lie properly, wrap six or seven passes of clothesline around the tank and the wagon bed, and that should be that.

IMPORTANT SAFETY CAUTION AND DISCLAIMER

CAUTION: *When you are moving the heater into place on the wagon, have help. The wagon will tend to tilt and roll as you lower the heater onto it. You can try to place a block in front of the rear wheels, to keep it from rolling forward, but that is a bit chancy. You'd better try it with two people.*

At this point in the book I must make one thing very clear. I cannot be responsible if you take my suggestions and, in the course of following them, you have an accident that damages your property or causes you injury.

I am giving you the benefit of many years of my own personal experience, suggesting shortcuts and techniques which have worked for me very well throughout my career. I strongly urge you to practice safety with tools, take precautions in lifting, be aware of your personal limitations, and observe the utmost caution in your demolition and construction activities. Use safety glasses, gloves, and protective garments whenever you are working around flame, cutting tools, and extremely hot or cold plumbing materials or objects. Do not dig in the ground so deep that your armpits are below the level of the excavation rim, without someone present to watch over you and what you are doing. If you are crouching down, digging in an excavation where your head is below the ground surface, be sure that the earth is not loose and crumbling. People have been buried, crouched over in ditches, and have suffocated before they could be dug out. Dig standing up, or hire a skilled laborer.

Don't climb shaky ladders. Don't put your weight on the ladder's fold-out shelf, or climb higher than the safety regulations allow for the ladder type and height. If you tend to have vertigo (dizziness) or a fear of heights, don't work high or in circumstances which bring on vertigo or fear syndrome.

With my warnings in mind, take great care in your moving of the water heater out of its location and to the disposal area. If you come to stairs and must move the heater up or down to its ultimate destination, get help. *Don't try to do it yourself.* One person can grasp the heater's feet and the other can take hold of the two spuds or nipples. Then you can carry it like a stretcher.

Replacing the heater is a matter of reversing the procedures I've outlined above.

If, in the course of removing your old heater, you have had to cut copper pipe, and you are not familiar with soldering techniques, please turn to Section Five where they are described in detail.

If you have galvanized-steel plumbing, installing your new hot-water heater is a simple matter of setting new nipples in the hot and cold tank tappings at the top of the heater. I usually use ¾″ nipples, 3″ long, which is normally expressed as ¾″ x 3″. Make sure that you use a proper amount of pipe-joint compound (pipe dope). On top of those, place two ¾″ galvanized unions. If your heater is of a different height than the old one, you either adjust by (a) shortening the existing risers, replacing them with nipples or cut-and-threaded pipe of the proper lengths, or (b) lengthening them by adding ordinary couplings and other nipples of the proper lengths. It's as simple as that.

HELPFUL HINT: If your present heater water shut-off valve has been dribbling all the time you've been doing your heater caper, now is the time to replace it. Shut off the house water main and change the leaking old hunk of junk then and there.

If you don't mind spending a few bucks more, put a valve on the hot side as well. *Recommendation:* Use American-made gate valves only. They're going to cost you quite a bit more than the foreign-made valves, but they're worth the difference, believe me.

SPECIAL COMMENT: When you're off buying your new water heater, be sure that you purchase a temperature-and-pressure-relief valve (T & P) at the same time. The code is quite clear on the subject. You must have a T & P on every water heater. It's protection for you.

If your heater's controls fail, and the water begins to boil, without a T & P on the heater, it could blow up and take a wall or two with it. Or you could turn on the hot tap and be scalded by live steam.

The T & P relieves the pressure and directs it to the floor drain or to the outside of the building. As soon as enough steam and water have "blown off" and cold water has taken their place, the unit is safe once more, until the steam or temperature build up to critical again, at which time the cycle will be repeated until you notice the problem either through the sounds or the extraordinarily hot water. If this happens, don't even try a repair. Just shut off the gas (or electricity, if it is an electric heater). After it has cooled, yank the heater and replace it.

It simply isn't feasible to wait weeks for a new control for your old heater. If the control is so old and feeble that it has "blown its top," chances are that the tank is not far behind.

You handle electric water heaters exactly as you do the gas models, except that you must be sure that the electricity to the heater circuit has been switched off at the breaker or fuse box. CAUTION: *Test at the terminals where the power comes into the heater, before you proceed further.*

HELPFUL HINT: There are small neon-lamp-type circuit testers available in any hardware store for less than $3. All that you need to do is touch the two leads of the tester to the two terminals on the heater's thermostatic control. If the neon lamp glows, go back to square one and check the circuit to the water heater. If you can't locate the proper breaker or fuse, turn off the main power switch or breaker to the house. Even with the house dark, and all power seemingly off, test the water-heater circuit again with your tester. Strange as it may seem, sometimes 220-volt appliances such as water heaters can be on special circuits that must be turned off separately. What you may think is the house main breaker or switch may have been bypassed to another 220-volt submain or "feed." *Don't take chances with electricity. If you are in doubt—don't!*

Once you're sure the power is off, disconnect your power cable, noting how the connections are made, and tape up the bare ends of the lines very carefully, taking many turns around each connecting wire, and then on up the cable, to make sure that no one can be accidentally shocked and perhaps electrocuted.

Once you've taped up the leads, it should be safe to restore power to the rest of your house. If you've localized the breaker or fuse for the water heater, naturally leave it switched off or unfused at the box, while you install your new heater.

The electricity is the last connection (on gas heaters, the gas is the last connection).

You should have absolutely no difficulty connecting the gas service to your new heater.

Nearly all domestic hot-water heaters have ½" gas inlets to the controls. Most "listed" or approved flexible connectors have a male "union end" and a female "union end." By "union," I mean that the two ends are designed to be separated from the plastic-covered connector and screwed into the control and onto the lever-handled gascock on the steel pipe, with a minimum of fuss and special fittings. Some of these connectors come with special gascocks, designed to go right into the connector female threads.

In any case, take my suggestion and replace the gascock and connector for your new heater and start the whole shooting match from a fresh beginning.

I haven't mentioned the vent in my discussion of the removal and replacement of a failed heater. My reason is simple. In most cases, you remove one section of the vent when you take the old heater out, and slide the new heater right under the vent to connect up. When you've connected the water and gas, simply replace the section of vent you removed before, snugging it up to the new heater's draft diverter.

SPECIAL COMMENTS: For years, domestic water heaters have

been fit up for 3″ vent pipe. Some jurisdictions now require water heaters to be vented into 4″ vents. Normally, if your old heater has a 3″ vent still in place, you can use the same vent for your new heater.

In new construction, however, it is best for you to check your local building department for the code vent size in your area. The *UPC* permits 3″ (a minimum of 7 square inches of vent-pipe cross-section area).

The draft diverter is that umbrella-like assembly that is placed over the heater-vent hole on top of the heater. It usually has three struts, or legs, which fit into three small slots in the sheet metal of the heater top. CAUTION: *Under no circumstances must you make your vent pipe tight to the heater-vent hole. You must attach the vent pipe to the top of the draft diverter and place the diverter properly, with its legs set firmly in their proper slots.*

The diverter is designed to prevent a gust of wind, coming down the vent pipe from the outside, from blowing out the pilot flame or the burner flame itself. The diverter also has a dish-like plate that prevents water from dropping into the firebox. The water strikes the plate and evaporates before it can enter the heater's venting system.

That dish-like plate diverts gusting wind, as well. So be careful where you lay that wonderful gadget. It's designed perfectly for the job it does. Put it where it belongs, between the heater and the vent going through the roof.

SPECIAL COMMENT: Some of you may become frustrated when you try to take loose vent pipes from furnaces, water heaters, and ranges. Some of them may seem to stick together at the joints and defy you to take them apart. When that happens, examine the joints carefully, using a flashlight. You'll probably find that the original installer had fastened the joints together with sheet-metal screws.

When you remove the screws, after you've fought the chore for awhile and before discovering the fasteners, you may find that you have deformed the pipe joints a bit. No matter, just get a pair of pliers and bend the ends into roundness again. *Have patience* with all sheet-metal products. *If they went together once, they'll go together again.*

If an adjustable (segmented) vent elbow has come apart, don't try to repair it; buy a new one.

If you have taken care in the installation of your new water heater, it should be good for a lot of years of trouble-free service, *if* you take a few precautions.

Every six months, blow the heater down by attaching a hose to the drain bibb at the bottom of the heater and opening the bibb wide. That will blow out all of the sediment and debris that may have collected on the inside of the tank and settled to the bottom.

Once a year, shut the heater down and attach your hose again. This time, shut off the cold-water supply to the heater. If you have a hot-water valve as well, leave it open. Then open all the hot-water taps in the immediate area, including at least one higher than the heater top, if that is possible. Then let the tank drain completely. Shut the drain bibb once more and fill the heater again, shutting off the hot taps in the house as they stop blowing air and you get water.

Immediately after you've refilled the hot-water system, blow down the heater immediately, as described in your every-six-month procedure above.

HELPFUL HINT: When you are emptying the heater, as the water-level lowers, open the T & P, by prying up the lever with your fingers until it stays open. That will let a great deal more air into the tank, allowing it to discharge water into the drain bibb at a much faster rate.

Important note: When you install your T & P, be sure that you run a full ¾" length of either copper or steel pipe to within 6" of the floor. If you have not yet studied Section Five's instructions on soldering copper, and you plan copper plumbing, hold off on your run from the T & P until you've read that section, which deals with the construction of plumbing systems.

As a matter of fact, if you have the time, and you are beginning to notice dampness around your water heater, you might as well get on with developing your soldering skills before they're needed urgently.

The secret to maintenance is prevention. The secret to repair is regular and sensible maintenance, which will familiarize you with your house systems.

SECTION 4

Rainwater Drains and Site Drainage

For the record, plumbers aren't supposed to get into solving problems of flood control or the design of rainwater systems. Those jobs are properly handled by architects and engineers rather than by wrench-pullers.

Over the years, however, I've found that when people get water in places where they don't want it, or if they want water taken to where they don't have it, they usually call the plumber.

In the beginning, I used to turn down flood work that didn't have to do with plumbing failures, and refer my customers to local professionals for advice.

In time, however, I began to notice that a lot of their successful solutions were based upon plain old down-home common sense. In a lot of cases, in my own day-to-day plumbing experience I flashed on situations that would have made my on-the-spot assessments and remedies every bit as effective as the ones proposed by the big shots downtown.

I don't have anything against architects and engineers. They're necessary, valuable members of the professional community. In the long run, where proper planning for construction of any kind is concerned, I strongly suggest professional consultation rather than hiring a builder or other craftsperson to both plan and build.

But, in the area of the management of your immediate environment—insofar as dewatering and runoff are concerned—unless the problem is one of fundamental house or building design, or drastic alteration in the land configuration (mudslides, creek or river diversion, or major natural disaster), an experienced plumber or knowledgeable builder can be quite helpful.

By extension, since this is a self-help book, you should be able to make some important observations on your own, leading to useful solutions of simple, straightforward problems.

Warning: The moment you find yourself in a questionable situation in which you cannot find a common-sense answer, stop what you're doing and consult a professional. Don't take chances with rain and storm water management problems. A wrong guess can mean someone's life.

GENERAL RULES OF RAINWATER MANAGEMENT

Every house built to any code in the nation must have provision for carrying off rainwater from the roof and paving.

Normally, a pitched roof drains to its eaves. A flat roof can drain either to its outer edges (which implies a slight pitch outward) or to drains on the roof surface itself.

The pitched roof disposes of rainwater runoff into eavestroughs or gutters, which deliver to sheet-metal or plastic leaders or downspouts running down the outside of the house.

The flat roof—and even some larger pitched roofs, with multiple peaks and valleys—deliver to internal drains of standard drainage materials, usually cast-iron soil pipe and galvanized-steel or copper pipe. Many codes prohibit the use of plastic pipe as rainwater conduits inside a building, unless it is underground, which usually implies a main in place, beneath a slab of concrete.

Your house sewer system and the rainwater system *must* be kept separate for very practical reasons. If you live in a town that has a storm-drainage system separate from its sewers, obviously you can't have sewage running in lines designed for relatively uncontaminated runoff. By the same token, it could be hazardous to the public health to overtax a sewer system with massive storm runoff that could result in main sewers backing up into dozens and even hundreds of homes.

Some towns have combination sewer-and-storm systems, but many of those that do require separate connections for house-sewer and storm lines.

If you're on a septic tank, the separation of systems is critical. Obviously, rainwater runoff could easily overwhelm even the most grossly overdesigned package.

Yet, with all of this logic, a prime cause of malfunctioning house sewers and rainwater systems has been do-it-yourself cross-connection by zealous home-craftspeople who want a nice, tidy rainwater system discharging into neat plastic-pipe hubs, leading to bootlegged ABS or PVC connections into the building sewer.

If you happen to be one of the people who has done this, don't wait too long to separate the systems; one day, this advice may save you hundreds of times the cost of this book. It could conserve your and your family's health and well-being as well. You only know about such things in emergencies. *Be prepared.*

ROOF DRAINAGE

If your house is old and a bit rundown, chances are your roof gutters are in worse shape than the rest of the place. The reason: Most people only think about their gutters during a rainstorm. The rest of the year, it's "Oh, I gotta get to those gutters someday, but *first* I gotta build the new barbecue."

Forget the barbecue, until your home's rainwater system is sound and working the way it should. It's a hell of a lot more important to have your gutters clean.

One of the biggest problems with gutters is the fact that they're high off the ground, and therefore a bit scary for the average nine-to-five office worker to repair or maintain.

It is important to understand the technique of using ladders properly, in order to minimize the fear. Wherever possible you should use a sturdy stepladder, capable of standing by itself, and you should have another person there to brace it while you are standing on it. If the roof is too high for your stepladder, and you require an extension ladder, it is even more important for you to have another person in attendance.

As you are going up, *never look down!* As you are coming down, feel each step with your foot as you lower yourself, keeping one foot and both hands solidly in place. If you must look down, focus on the step-bars only. As much as possible *avoid looking down to the ground or to the person below.*

When you get used to the ladder's "feel," you should be able to bend that rule a bit, without too many ill effects.

As you are going up or coming down, most ladders will quiver and deflect a bit. It's the nature of ladders. The longer they are, the more vibration and sag you will sense. *Be assured:* If you are of reasonable weight (within the range of the safe-weight limit on the OSHA sticker affixed to the ladder frame) you should be okay, going both ways. I weigh *over* 200 lbs., yet I have no fear of going up an eighteen or twenty footer which is OSHA-rated at 200 lbs. The important thing to realize is that, if you keep your head and don't panic, and if the base of the ladder is secure in the hands of a friend, or tied off, you should have absolutely no problem working near the top of a ladder of any size.

HELPFUL HINT: If you have become used to your ladder and have had a friend securing it for you for awhile, it should be okay for you to go "solo." In order to have more security than just setting the ladder feet firmly on the ground, it's a good idea to run a "cinch line" down the side of the house or building. The cinch line should be fastened at both ends to a solid anchorpoint (pipe, beam, doorframe, etc.). If necessary, bolt or nail a two-by-four in place and secure to that.

Now, with your cinch line secure at both ends and taut, as you move your ladder down the side of the house, working, take another line from your cinch line to the feet of your ladder. The way you do that is to loop your short rope around the cinch line and tie each of the two free ends to each leg. It's simple, much safer than unsecured ladder feet, and very reassuring to the person doing such work for the first time, without help.

CAUTION: *Do not* ever *stand on the top three steps of a step ladder, including the top platform. Do not* ever *stand on an extension ladder where you cannot get a firm grip on the ladder or upon a portion of the building itself. Do not* ever *erect an extension ladder at a shallower angle than that recommended. If in doubt ask the clerk at the store at which you bought it.*

FURTHER CAUTION: *If you tend to have vertigo (dizziness) or other disorienting ailment or tendency, don't climb ladders of any kind, under any circumstances, unless you are under the direction of a public safety officer or fireman at the scene of a fire or other disaster.*

You are more likely to have a ladder accident inside your house, hanging up Grandma's picture, than you are twelve feet off the ground, cleaning your house gutters. Why? Because, in the house, the height is not great, and you're used to the spaces in which you're working. On an extension ladder up near the roofline, you're going to be conscious of every move you make. A little fear goes a long way in safety.

At least I've got you up on your roof now. While you're there, have a look at the entire rainwater drainage system.

If it's a pitched roof, are your gutters clear? Are the leaders or downspouts firmly locked into them? Are they free of leaves, twigs, bits of gravel, birds' nests, hornet houses, and various balls and other kids' toys lobbed from the ground by the local scamps? Chances are that at least one of the obstructions I've suggested is lodged in one or another crucial place.

Long-neglected gutters have a tendency to build up with a silt-like accumulation of metal oxides, wetted dust, decaying shingle wood, asphalt particles from composition shingles, and any number of other environmental particles. In ten years of neglect, it's possible to find a forty-foot-long window box, complete with growing and crawling things. In that case, replacement would be in order.

If you're like most other people, you've had someone up there, or have been up yourself, at least once every two or three years, so your chore will probably be a dirty cleanup job, which can include a possible encounter with a critter who resents a disturbed environment. If that critter happens to be a hornet or a wasp, and you're fourteen feet above ground, you could be in for a problem, a dangerous stinging injury.

HELPFUL HINT: *Always* carry an aerosol of wasp-and-hornet control, when you work high. Spray the general area from below the gutter and watch for results *before* you commence work. This stuff works on bees, spiders, and other noxious insects as well.

CAUTION: *Whenever you work with a person below you, be abso-*

lutely sure that any tools you carry aloft with you are secure. I usually carry tools in an apron belt and I fasten thongs to screw eyes in hammer handles, to handle holes in wrenches, and to drillings I've made into other tools. It is very important that the person working below be equipped with a hard hat. They are relatively inexpensive and can be purchased at almost any hardware or building-materials store.

Suppose you are stung while aloft on a ladder. Your first instinct is to try to swipe the insect off your arm or face. In many cases, *if you do, you're in serious trouble!* Your human instinct will almost force a reaction to any kind of stinging injury, *but* if you are mentally prepared for what may happen while you're on a ladder, you can overcome that reaction. The important thing to remember is that most insects will go away, after they've stung you. You must concentrate upon getting down the ladder as soon as possible, especially if you've been stung in the face. You could be allergic and your face could swell to the point where you can't see, in a matter of minutes, or you could go into dangerous anaphylactic shock.

The most important thing to concentrate upon is getting down as quickly and as carefully as possible, considering the circumstances.

Even if you've received multiple stings, it's the safe getting-down that's important, and you probably won't make it if you don't keep cool, even with the pain, misery, and fear.

Take a firm grip on the ladder rails, and start your descent, slowly and deliberately, keeping as close to the ladder as possible, so that your body sort of slides over the step-bars. If you keep calm and move with a deliberate rhythm, you will probably not be stung further. Insects react to panic with fright and defensiveness, just like human beings. Calm vibes usually mean that they will turn their attentions elsewhere. They'd much rather deal with things not as monstrous and alarming as a five- to six-foot giant.

The best implements I've yet found for cleaning gutter systems are an ordinary serving spoon with or without drain holes, a hammer, fasteners to resecure sagging portions of the gutter, and a kid's sandbucket, hanging from my belt on a short cord, to receive the dreck I dig out of the troughs.

Most of the loose material, leaves, twigs, and the like are easy to remove with bare hands. Pine needles and other tightly bound vegetation pose a bit more difficult problem, especially where the braces for metal gutters occur. Some types of braces run across the tops of the gutters. Other working obstructions include the ferrule-type nails that pass through the gutter proper, and the ends which seem to accumulate the most difficult-to-remove materials.

In the course of your cleanup, if you discover "soft spots" in the bottoms of either metal or wooden gutters, you are looking at the symptoms of general rusting or decay. The remedy: Replacement before the next rainy season. Don't wait too long! Before you know it, the winter storms will be upon you, and it will be too late for another year.

This is probably the only section of the book where I'll flatly tell you to hire a job done if you have the slightest hesitation about doing it yourself.

Gutter replacement usually has more hazard and less glamour attached to it than any plumbing job (toilet clearing and replacement excepted).

The tendency is to put it off until a "tomorrow" that never seems to come. That should not be! A failed gutter system can have catastrophic consequences. Direct fall, wind pressure, or capillary attraction can cause water to run down the sides of your house, enter cracks or holes, get under loose or weathered siding and begin seriously rotting the siding and the area between the walls. That same water, either falling straight to the ground from the eaves, or streaming down the siding, can attack and undermine your house's foundations.

Roof-drain blockages may lead to water buildup, which can cause roofing-surface failure, buckling, and even collapse.

So here you are, between a rock and a hard place. On one hand, it's hazardous to sip a can of beer on a ladder, doing a dirty, aggravating chore, but on the other, it may be worse to have to strip the windward walls of your house or shore up a sagging foundation, half a dozen years from now.

If you haven't got the stomach for the job, get it done by a pro, and sit in front of the TV. By the way, make sure that there's enough money in the bank to cover the check you'll have to write.

If you want to get in "good shape," acquire skills and the discipline that will help you later in serious plumbing repair and construction, there's absolutely nothing better than gutter and roof-drain work.

You will notice that, in time, your "ladder etiquette" will improve, along with other skills. Healthy fear will always be present, but the quaking panic will miraculously vanish.

If you can clean the gutters from the roof, good. It will take less time and give you a better overall view, as well as easier access. CAUTION: *If the roof is moderately pitched, work with a safety line secured around your chest with a square knot to either a solid masonry or brick chimney, or a metal plumbing vent pipe (plastic won't do for a number of reasons). Make sure that you keep tension on the line at all times. Take care that the rope does not develop too much slack as you move from place to place or you could fall the length of the slack. Crunch!*

If the roof has a heavy pitch, work from the ladder.

ADDITIONAL CAUTION: *Be careful when you walk on a roof, especially if yours is a very old house. Always test as you walk; put one foot forward and try your weight as you go. One foot is always in a place on which you've already put your weight.*

It's like walking in a "mine field," and you should treat it similarly. Remember where you've walked, and try to retrace your steps when you return to your ladder or other roof-access point. Of course, if you *know* that your roof is completely sound, from personal experience, there's no problem.

The things to look for in roof drains are obvious. If the screens have come off, or if they're clogged with debris, use straightforward, direct remedies.

It is important that the screens be in place, for obvious reasons. If they've been removed or blown away, the usual source of supply is the sheet-metal shop, your local building-materials supplier, or the shop of a heating and ventilating contractor. Roofers and roofing-supply stores are also fine sources.

While we're looking at the drain-and-gutter system of your house, let's take a look at the discharge ends of your leaders or downspouts. Do they just come down the side of the house and drop the water right next to your foundations? If they do, you're in trouble.

They should be delivering their flows to some place away from your house that will either drain or absorb the runoff. Modern homes in well-developed urban complexes have a storm-drain line that carries runoff to the storm-drain mains in the streets.

If your area has a storm-drain system, and your home was built before the system was constructed, you may have an unused resource close at hand.

HELPFUL HINT: Give the town engineer, local utility district, or building department a call and find out if you can hook up to the storm-drain system.

If you get a go-ahead, you'll have to endure the same song-and-dance as you do when you want to get a permit to plumb your house. You'll probably have to make a drawing and submit a proposal. Later in this section, I'll walk you through the entire process.

Back to our inspection. If you already have a disposal network for rainwater, usually cast-iron or plastic hubs placed to receive the discharge from the leaders or downspouts, make sure that they are properly connected and functioning. That usually means loosening the leaders a bit so that you can pull them out of the hubs. You'll be surprised what you'll find lodged in those hubs. Would you believe dead rats, snakes, and various insect nests? Often you find things in leaders and drains that you can't possibly imagine would fit into them. How would a two-foot stick get into a drain with bends and other impediments? I don't know, but I've found them.

HELPFUL HINT: Take one or two leaders per weekend, and in a couple of months' time you'll have covered the entire system. When I say covered, I mean cleared, cleaned, reattached, and tested with a hose.

If you find blockages in the storm-drain lines in the ground, you'll have to treat them as you would sewers (please read Section Three).

Warning: If, in the process of checking your rainwater drainage system, you discover that it is cross-connected to your sewers in the house, you may have to separate the two systems (See Section Five on construction).

HELPFUL HINT: If the two systems are interconnected, and they have been satisfactory throughout the years, check with the building department and find out if such interconnection is allowed by the local code. *More importantly,* find out from the language of the code how such systems are to be connected. Some codes permit the interconnection of the two systems a certain prescribed distance beyond the house foundations, where smaller sewer and house rainwater-drain lines connect to a larger main going to the city service. Other local codes don't seem to give a damn how the two systems are connected to each other, or where.

The important thing here is to find out what must be done to insure the proper functions of both your sewage and rainwater disposal systems.

I'm not trying to make waves. I just want you to have the best systems possible in the circumstances. Knowledge of what should be—as opposed to what you have—gives you options.

If your rainwater system is discharged to the surface (ground level), you've got to be prepared to change it. If your roof-drainage is running willy-nilly, you must do something about that soon, as well.

Your leaders must not discharge to your foundations.

The first thing you must do is extend the discharge elbow of the leader a bit to get the water to an adequate drainage surface, like a properly pitched walkway or patio. By properly pitched, I mean that the concrete or masonry has been built to drain the water away from the house. If it doesn't do the job, then you're going to have to take the water farther out to some place which will move the runoff to a drain or percolation area.

HELPFUL HINT: Some folks I know buy polyethylene or vinyl flexible tubing that is designed for this kind of work, and attach it at the beginning of the rainy season. When the season ends, they take it off, roll it up, and store it for the next year. The stuff is very thin (around eighty millimeters thick). Actually it looks like a tubular plastic garbage bag, only the walls are a bit thicker. It can be tied around your leaders and rolled out to a good drainage field or slab.

CAUTION: *Don't use this tubing in areas that are subject to heavy foot traffic. It doesn't take much of an impact to breach the plastic, especially when it's running full of rainwater.*

Another material people are finding useful in these kinds of situations is Hancor polypropylene corrugated plastic conduit, some of which is perforated for irrigation, leach-field, or drainage applications. The type which could be used as a drain pipe is *not* perforated. It comes with a limited range of fittings, tees, elbows, couplings, and adapters, *but they are not completely watertight and cannot be made watertight easily.* The pipe is made in coils of 250 ft., and is commonly available in sizes from 3″ to 6″. Larger sizes may be made, but I haven't seen it around. This conduit or pipe may be laid on top of the ground as a temporary seasonal remedy for your rainwater runoff problem, or it may be buried and used as a permanent storm-drainage line. CAUTION: *This material is not UPC Listed in the 1979 edition, and you are warned that, if you bury it, and use the fittings which are designed for it, you run the risk of discharging significant amounts of rainwater at each fitting or joint.*

Some people have achieved seals at the fittings by pouring them solid in concrete. Others have used filler-type cements and glues between the fittings and the pipe ends to be joined.

Whatever you do, be careful. Remember that this material, being non-code, may have to be dug up later, as a result of a building inspection.

For temporary surface use, you can't beat it. It doesn't damage easily, although it may trip Grandma, if she happens to be walking around the yard during a storm, and Uncle Elmer's wheelchair just won't get over it, when he's pushing his wheels lickety-split for the kitchen door to get in out of the rain.

Yet, all things considered, that Hancor pipe may come in handy for other surface emergencies, which I'll discuss in just a bit.

For buried storm-drains, stick with ABS and PVC schedule 40 pipe and fittings. If you want to build such a system, flip a few pages to Section Five which will give you the necessary details.

I mentioned before that, if possible, you should conduct the rainwater runoff from your leaders or downspouts away from the foundations of your house through temporary tubes which I've just finished describing. You will be saved the trouble of constructing such a temporary network if you have been prudent in the past, and have built concrete walkways, patios, and driveways with drainage in mind.

If you've bought into a "can of worms" in which the prior owner has done a misdirected do-it-yourself number on the place wherever you look, and all the concrete around your home slopes the wrong way, make up your mind to change that this summer.

It is far better to get that straight than to start piping runoff here and there, in an attempt to compensate for the former owner's lack of foresight (translation, stupidity).

Of course, I've assumed that the poor drainage is the fault of a non-contractor, because pros are supposed to know better. That's the reason you pay them. You buy the knowledge and experience that's supposed to go with their licenses.

Well, folks, sometimes it ain't necessarily so. Like the local contractor, who is the brother-in-law of a license-board member, who can't read a set of plans or add a column of three-digit numbers to save his life.

That contractor's name happens to be Elmer too.

Sometimes I think that his head needs a wheelchair.

I recently saw a $135,000.00 house that Elmer built. He should call his outfit the "Topsy Construction Company." Not a wall is plumb or square, not a floor level, and every piece of concrete on the property is pitched wrong or sprouting shear cracks.

CAUTION: *When you get serious about buying a piece of property, hose down all the concrete and see what the drainage patterns are. If you can, run a hose in all the gutters to check the rainwater drainage system.*

It blows me clean away when I see supposedly intelligent people acting like "tire kickers" when they go house hunting. They cluck and shake their heads when they see damp sills or water marks on ceilings, and when the seller tells them that the roof is new or the leak's been fixed, they smile and go on tapping walls and turning faucet handles as if they knew what they were doing.

HELPFUL HINT: National Plumbing Information Service, Post Office Box 3455, San Rafael, CA 94912-3455, has a *Home Buyer's and Owner's Checklist* available. It costs $5 and it covers everything: roof, finish carpentry, structural soundness, plumbing, conformity to electrical code, concrete, the works. It tells you what to look for and about how much it should cost to remedy construction defects or the ravages of age and use.

The *only* way to correct badly laid concrete is to break it out and start from scratch.

You can hire out the jackhammering or go to an equipment rental place and rent a compressor, pneumatic hammer, and points. If the job is relatively small, perhaps an electric hammer will be enough.

HELPFUL HINT: Sometimes it is sufficient to break out a 1-ft. to 2-ft. strip parallel to the house foundation and create a roll-edged trough or gutter for runoff.

Another method of disposal is into dry wells, when soil-percolation conditions are exceptional, and you have sufficient land. *For this remedy, consult a civil engineer.*

If a dry well is not feasible, you may be faced with digging trenches and hauling one whale of a lot of rock.

SITE DRAINAGE

If you have ever experienced a flooded cellar, basement, or garage, you know firsthand what it feels like to be a flood victim.

There is nothing to be done if your entire neighborhood, tract, or farm is under four feet of water from a swollen river or creek that has overflowed its banks. There's no place to pump or drain off the flood water. But if your house or building is situated in such a way as to be susceptible to periodic drainage problems or flooding when neighbors 100 feet away are high and dry, there are solutions.

CAUTION: *It would be best to get engineering advice and plans, if the problem is massive and regular, during every rainy season.*

If the problem is not overwhelming and occurs sporadically, during the worst storms only, the solution can be as simple as a small submersible sump pump and 100 ft. of 1¼" PVC pipe.

The first thing to be done in dewatering or site drainage is to establish "low points," places that puddle or fill up first.

Next, observe the patterns of water movement, as the runoff builds up. Often, it will channel itself into long low-lying trenches or it will move down slopes, over a broad front.

If it settles into gullies, you can use the simple sump-pump system. If water moves downward, over a broad front, it will probably require a long trench, dug across the water "front" or path. The ditch then drains into a sump, and the water is pumped to an established disposal area or storm sewer.

(If your land is flat, and your house or outbuildings are in a hollow, the modern version of an old-fashioned moat may be in the cards.)

I prefer the automatic, submersible-type sump pump, which operates efficiently and almost soundlessly, totally submerged in water. The least expensive of these, with a capacity of 2,000 gallons per hour or more, cost as little as $110. Some "all-bronze" 4,000 gph units can cost as much as $400, but I'd rather use two 2,000 gph pumps at $220 for the pair. That way, you have some pumping capacity left, if one goes out of service. As a matter of fact, I have hooked up batteries of three and more of those pumps for some major dewatering, and they've done one hell of a job.

CAUTION: *If you hook up more than one pump, be sure that you have a gate valve and a vertical check valve on each pump discharge. If you don't, it will be difficult to operate the system whenever you must remove one pump for periodic maintenance and replacement. Without the check valve, one pump could pump right back through another and nothing would come out the end of the pipe.*

In any case, you should have a check valve on any sump pump, to prevent backflow from the line, when it stops pumping.

The types of pumps to which I am referring generally discharge into lines of from 1¼″ to 2″. The 1¼″ pump is the most popular. As a rule of thumb, if you use one 1¼″ pump, 1¼″ to 1½″ discharge pipe is recommended, depending upon the distance the water must travel to the disposal area or storm drain. Two pumps use 1½″ to 2″. Two inches will usually handle three pumps as well. Four to six pumps are a comfortable load for 3″ pipe. CAUTION: *If your job requires more discharge capacity than that, you'd better see a civil engineer, and have a system designed for the job.*

Okay now, you have found a low spot somewhere, and you find that a great deal of the surface water that you want to control flows to that low spot. The next step is to build a sump. The best materials I've yet found to line the hole are two 4″ x 8″ x 16″ partition-type cinder blocks for the base of the sump. First, dig your hole over which you place a 14″ Sonotube, in an excavation 20″ in diameter (if round) or 18″ square.

(Sonotube is a fiber tube made for heavy construction. You've probably seen them used as forms for concrete columns of high-rises. They are normally stocked in 20-ft. lengths, but most building-materials supply yards will cut them to almost any length you need.) Then you pour five-sack concrete "wild" (one part cement, to four parts aggregates and sand) between the excavation and the Sonotube. The whole process is practically self-explanatory.

HELPFUL HINTS: The partition-type cinder blocks are half the thickness of regular cinder blocks, which are 8″ x 8″ x 16″. So, if you can't find the halfblocks (partition-type), just dig your hole 4″ deeper and use ordinary cinder blocks.

Normally your sump would be 18″ deep. When the concrete has cured, you can either cut out the Sonotube, which will leave a nice concrete sump, or leave the tube in place and coat it with epoxy paint so that it will stay intact over the years.

To be sure of a solid, smooth, homogenous concrete wall, you must take care to "shake down" your pour by tamping it through-and-through with a rod or stout stick.

The reason you make your sump 18″ deep is so that your pump's automatic diaphragm switch will work properly. Usually the water must be over the top of the pump, before enough pressure is brought to bear upon the switch and the unit is turned on. Pumps range in height from around 12″ to around 16″. The 18″ sump gives plenty of allowance for proper switch activation.

When I say five-sack concrete, I usually mean four to one. One sack of Portland cement to four sacks of sand and gravel, or less than the sack quantities, in proportion, to your pour requirements. When you pour it "wild," you pour it to fill the uneven cavity formed by the excavation and the Sonotube. This is differentiated from form-pouring, where you form the sump walls with form lumber and pour to the form. With Sonotube, the inner form of the sump is the Sonotube itself, and the outer form is your excavation. Easy? You bet!

When you've allowed your sump to "cure" properly, it is best to plant the area for a few feet around your sump with any sort of persistent, erosion-resisting vegetation. You don't want the surface water to bring too much mud into your sump, or you'll be swamping it out every day during the worst of the rainy season, and that can be an awful drag.

HELPFUL HINTS: It is best to put a coarse screen or grate over the sump, to keep the worst of the flotsam and jetsam from dropping into it. A super material for this kind of work is aluminum expanded metal, but make sure that you get a sufficiently beefy piece, or you'll have it caving in in the spring, when the gang starts baseball practice or scrimmaging.

Getting power to your pump is another thing again. Consult your local electrical code.

HELPFUL HINTS: You will probably need a separate breaker for each pump station. If you are taking the power quite a way out from the house main, you will have to consider wire size as related to distance, which means that you'll need to be very concerned about those points when you are studying the code.

Direct-burial cable comes in all gauge sizes. That's the kind of wire you'll probably be running to your pump in a field. Make sure that you check your local code to determine whether or not that kind of wire is acceptable. If it's not, then you'll probably have to run conduit. The easiest to use is PVC, which most codes allow. Be sure you understand the kinds of receptacles that are required. In most jurisdictions, waterproof (Bell-type) boxes are specified. Ground-fault interrupter receptacles are also common requirements.

Frankly, if you take my advice, let an electrician hassle with the power to your pumps, unless you are very comfortable with it, and have done your homework. CAUTION: *Make sure you check your pump's voltage and amperage specifications. I've seen otherwise meticulous craftspersons hook up 220-volt motors to 110-volt services and the other way around. Don't take anyone's word for what you have bought, as far as power is concerned. Don't even take the information on the outside of the box as Gospel. Look at the U.L. plate on the motor and read the numbers engraved on it. Then you'll know for sure.*

I've seen factory-packaged pumps with the wrong information stamped on the outsides of their packing cartons, especially foreign-made pumps.

If your pump is installed in or close to the house, of course you can run the code-prescribed Romex or "three-wire pull" in conduit.

Switchgear is as prescribed in the code. Some codes require a local breaker or disconnect (switch), which means that your pump site may have to include a small doghouse with these hunks of electrical "jewelry" installed.

If you're out in the middle of nowhere, of course, getting the power to the job and hooking up will probably be rather simple and uncomplicated.

CAUTION: *If you run direct-burial cable, be damned sure that it's down in the ground below plow or digging level, and that it doesn't run across or near pipelines or other utilities that require frequent excavation. Some folks I know mark the line of their cable with a small decorative fence, a line of hedge, or some other distinctive and permanent reference. The same caution applies to the pump-discharge line, which should be buried deep enough to avoid damage in the normal course of running your home or farm. Standard burial depths for both pipe and cable are 6" to 12" in areas that are not subject to plowing or frequent excavation. That means 6" to 12" of soil* on top *of the pipe or cable.*

If you're having water runoff down the face of a slope, the best way to control it is to dig a trench, which should be as wide as the width of the scoops on one of those *Ditch-Witch*-type ditch-digging machines (8" to 12"), which you should be able to rent from most equipment rental yards. They have a revolving chain "stinger" on a short boom, which does a great job of digging down to ten inches or so.

You'll see me refer to that wonderful little machine plenty in Section Five.

The Ditch-Witch and other little handlebar-guided ditch-digging machines of the same type are the fastest and most efficient way to make a uniform ditch. They permit some degree of precision with respect to the slope you must build into your trench. The slope is essential to moving the water down the trench to the sump. The pump does the rest.

The length of your excavation will be pretty much determined by the width of the water "front" during the period of greatest flow. If the water flows down a hillock which is twenty yards wide at the top and sixty yards wide at the bottom, you should consider the necessity of intercepting the flow near the base of the grade, with a trench almost sixty yards long. On the other hand, if you have a place where two hills meet, rather high up on a slope, and most of the water flow comes to that point, perhaps a rather short trench will do, just below the meeting place of the two hills. The advantage of trenching up high is the nice fall you have to your sump pit.

Speaking of "fall," there is a plumbing rule for all drainage. It is that you should slope your draining surfaces one inch for each four feet of pipe or trench to the low point at which you locate your sump pit or catchbasin. If you have 30 yards of trench (90 ft.) the slope from the high to the low end should be around 23". In most cases, given the type of equipment you are forced to use for this work, it's damn near impossible to get that much fall, even if you start a couple of inches below the surface at the high end. The rule allows you to have that fall to 11½" in places where it's impossible for reasons of topography, soil composition (boulders, heavy clay, hard pan, rock and gravel, and other obstructions at prohibitive levels), and available excavation equipment.

Obviously, you would be hard pressed to get 24" of fall in a trench for which you have a machine-capability of only 10" to a foot.

But then, there's always the "Siberian teaspoon" (an ordinary shovel), or an "Ethiopian toothpick" (pick mattock), and your two good arms (If they aren't good now, they sure will be, by the time you're finished).

The moat-type trench is usually a trench above your house on the hillside to the rear. It is cut just a bit wider than your house, plus about ten feet on each side. Then two trenches are dug (one on each side of the house), that join each end of the rear trench to form the letter "U," upside-down. The lateral trench above catches the water, and it flows down the two side trenches, each of which can terminate in a sump-pump station. Or, you can dig a pair of angled trenches that join at an apex in which you can locate one pump station, below your house.

One of the biggest problems with trench networks is keeping them open and operational. Almost as large a pain is the simple hazard to human life and limb posed by these open ditches about the property.

The answer is that you must fill them with rock, and the only satisfactory rock I've ever found for this kind of work is ¾" river-run rock, or ¾" crushed rock. If you get smaller sizes, you may have some difficulty later when the tighter-packed gravel begins to fill up with silt and other debris, and stops passing water.

Where you have very fine, loamy soil, it is best to line your trenches with agricultural polyethylene or vinyl film, twelve mils (thousandths of an inch) thick or thicker.

The film acts as a kind of waterproof conduit, and the rock holds the trench open, allowing the water to drain quickly and thoroughly at each barrier.

If you have high flow rates, increase the number of lateral trenches on the hillside behind the building.

If your home is set low, surrounded by high ground, you are almost forced to dig a trench around the entire house, except in those places that you know from experience have no flooding problem. Wherever you break the continuity of your trenches, you must have a sump pump. If all trenches fall to one point, one pump is sufficient.

Remember that Hancor polypropylene tubing I mentioned earlier? Well, here's a wonderful place to recall its properties: It's tough, it's durable, and it's made in long coils. Also, it can be purchased as a perforated tubing. If you put a bit of rock in the bottom of your trenches, then lay perforated Hancor on it, and fill the rest of the trench to the top with rock, you won't need the vinyl or polyethylene trench lining, and you'll get amazing flow through the corrugated tubing. Another advantage of the Hancor tubing is that you can get those handy easily installed fittings, and in this type of installation it's an advantage that they leak, because they will accept water coming into the trench.

SPECIAL COMMENT: If you have a subterranean water problem (underground water flow), and you want to keep it from coming into your cellar, dig your trench to intercept the flow, and let it come into your trench and your Hancor pipe, underground. Naturally, you don't line this kind of trench either. An advantage of this setup is that you can put a plastic-film "cover" on top of a trench which has only been partially filled with ¾" rock, put topsoil on top of that, and plant it. This way, nobody even knows that you have this wonderful dewatering network under the lawn. Of course, every time you water the lawn, part of your water will be carried off, but that's okay. You can put a standpipe on your sump pump and use the runoff for watering Aunt Phoebe's petunias.

If you get water in the cellar or basement, after placing your interception trenches above the house, you may have artesian or other subterranean water sources beneath the building. They're found all the time in the most unlikely places.

If that's your problem, drill a few test holes in your basement or cellar floor and try to determine the direction of flow and the place of greatest volume. Then break out a band of concrete about 12" wide, and as long as the basement or cellar is wide. Dig a trench, install your Hancor in its rock bed, and put in a sump and its pump at the low end of the trench. Cover with your plastic membrane, and pave over the lot. Even if you have people living in a basement apartment, they probably won't notice the sump pump cutting in, since these types of pumps are relatively silent under water. You can put a tight-fitting lid on the pit, and that should mute it entirely.

If you intend to dewater by this method, you should get a permit, since you're dealing with a situation that could affect the health and safety of your family, and could profoundly influence the future value of your home. You want it on the record as having been done (and, I hope, done well), and you will certainly benefit from your building inspector's observations and advice.

GETTING YOUR PERMIT FOR SITE-DRAINAGE CONSTRUCTION

If you have the survey drawings and/or plot plan of your house and property, you are ahead of the game. If you don't, you'll have to go down to the Hall of Records and have the clerk find you a copy of the survey map of your property. Chances are your house will be shown in relation to the metes-and-bounds. If it's not, no matter; at least you have the official description and boundaries of your property.

Often, if improvements have been made over the years, with permits, the local building department has an old application with a plot-plan sufficiently current to satisfy the inspection people. Just have them copy it.

If all you have are the boundaries of your property, with the survey data, just copy it down, in a larger scale, measure the boundaries with a tape measure (an accommodating family member or friend would certainly come in handy at this time) and then show your house and outbuildings, in scale, within the boundaries you've drawn.

Then show the locations of proposed trenches, pumps, and both the electrical conduit or direct-burial cable and your pump-discharge lines.

Next, describe your excavations—how deep you plan to dig your trenches, what you intend to place in them, and where they will terminate.

Give the technical data on your pumps, the manufacturer's designations of all electrical switchgear, pipe, valves, and fittings. You can meet the requirement by enclosing the manufacturer's technical bulletins or brochures.

If you are "cutting" concrete, or breaching any existing structural members in the house or in its immediate vicinity, describe what you intend to do, and if you have the ability, try to make a drawing illustrating that work, so that the inspector will know what you took your permit to build, how you intend it to look, and whether what you told him you were going to do matches what you actually built.

If you have no skill in the drawing department, just write out the best description you can and take that along with your plot plan, your brochures, or technical descriptions, and cross your fingers.

Don't be disappointed if the people down at the town building department smirk a bit and "tolerate" you. That happens often with do-it-yourself "project managers." Just be patient and ask as many "innocent" questions as you can. Bureaucrats love nothing better than showing off what they know. Appropriate "Oh, I didn't know thats" interspersed with a humble "Thank you" and "Gee, you've really been of great help to me," go a long way with these people.

They're just being human; everyone likes being needed. Applicants often have a sour attitude when they deal with regulators of any kind. And, quite frequently, chip-on-the-shoulder, dour-faced permit applicants ask for trouble when they demand this or that from bureaucrats who face the public for about 2000 hours a year, and have had it up to here with people who haven't got their acts together.

That's why it's so important for you to get your proposals straight and have solid answers prepared for anticipated relevant questions. Above all, treat the permit people and inspectors as you would any other neighbor, with courtesy. With a pleasant give-and-take attitude and good preparation, I guarantee you that even the gruffest, most businesslike and aloof public official will inwardly sigh with relief and try to see reason. That's more than half the battle won!

CAUTION: *If you're having trouble with your checkbook, don't write a check. It is better to pay with cash or money order. You'd be surprised how many people accidentally bounce checks on city hall. Believe me, nothing could be worse. It could translate into inspection nightmares later on, including revocation of your permit or assessment of outrageous penalties. The problem is, with a rubber check in their hot little hands, they've got you; they know where you are, what you're doing, and that you can't do it without their permission. When in doubt, don't write it!*

POSTSCRIPT

In closing this section, let me suggest that you read Section Five carefully if you plan gutter replacement or sump-pump installation. Construction procedures and rules are pretty standard for all classes of piping installations.

There is a hard-and-fast rule I always apply to plumbing design and construction: "The simpler the design, the more efficient the system, and the better the job."

Think before you install mechanical equipment, like pumps and controls. Can you do without them?

For example: Your hillside interception trenches could simply drain into Hancor *unperforated* tubing (installed underground) and be delivered directly to a storm drain.

Use your head, make the job as uncomplicated as you can, and do it with style and a sure hand.

SECTION 5

Plumbing Construction

The moment you decide to replace a substantial portion of your home's plumbing system, you have committed yourself to a course of action which, once started, must be carried to a *successful* conclusion.

There cannot be any "failure" in this kind of work. If you're a fair-weather tinkerer, don't even start. Keep your tools neatly laid out on your workbench in all of their shiny, pristine glory, but don't bother to pick them up and cut, wrench, or install a fitting or a single bit of pipe, unless you mean to persevere to the bitter end.

Your personal defeat, which can only occur at your own hand, will ordinarily require you to call upon professional assistance, and that can be as costly and humiliating an act as you will ever commit.

So, think long and carefully before you make your final decision, dear friend.

If, after this solemn warning, you are still determined to proceed, your first order of business is

PLANNING

If you are going to replace both your DWV (Drainage, Waste, and Vent) and potable water systems, the first order of business should be the DWV. Reason: It's relatively easier to work "around" an existing sewer (large pipe) than it is to work around water supplies (small pipe). This is especially true with copper water pipe.

An example would be a case where a craftsperson had installed all of the water pipe first and had created a network of hot and cold lines beneath the house and in the walls. Trying to get the much larger sewer pipes and fittings in place later puts you at the mercy of your own new water-supply installation. *It's very difficult to fit sewers around criss-crossing small piping.* It's relatively easy to work ¾" and ½" copper around 3" and 4" iron or plastic sewers. You have more options.

If you must decide between sewers and your water supply, you must determine the priorities. A decision between two ancient and leaking systems would lead to a sensible conclusion that your first priority must be the replacement of your water system. Its failure can lead to the more drastic consequences.

Your sewers are not under high pressure, and leaks can be controlled (see Section Three).

Severely leaking galvanized-steel water lines, under forty or more pounds pressure, can turn your house into the Okefenokee swamp in a minute, if something important blows.

By this time you should have become familiar with your house's plumbing system, through my suggested preventive maintenance and repair programs (again, see Section Three). If you feel that your water lines will hold for a year or so, and if your sewers are full of "bananas" (sags) and half-blocked fittings from the imprudent dumping of hard-setting materials like solutions of paint, plaster, cement, and other cohesive particulates that have settled into bends, then, the nod should go to a new sewer, hands down.

Since most initial replacements are water systems, I shall deal with that first.

COPPER REPLACEMENT OF AN EXISTING WATER SYSTEM

Once you have decided upon your project, you must continue your planning. Now, more than ever, forethought can save both time and money.

The considerations are: *Design:* How extensive will the system be?

Are you going to replace all of the water piping or just the portions which appear to be falling apart? *Structure:* Do you intend to go into the walls with your pipe? *Permit:* Will the job be large and complex enough to require you to file? *Finances:* Have you sufficient funds to complete the undertaking? Using credit for this kind of improvement requires a consultation with your banker or adviser. Usually, before you can be adequately prepared to present a good case for consideration by a lender, you must go through the motions, as they say. You must be sure that you give your application the solid feel of having covered all the bases. Financial people don't deal in the realities of flooded basements and stopped-up drains, *unless* they can see a way to make a *secure* buck out of the deal. That's only natural. They're in the business of making money for their institutions.

It is, therefore, your job to present a package that will support the credit or loan.

That means producing a workable plan and convincing the institution that you have sufficient expertise to carry off the whole scheme.

First things first!

You have a bad leak beneath the kitchen sink, and you have satisfied yourself that it isn't the flexible supply tubing that runs from the angle stops to the faucet. It appears to be in the walls. You may have located a rusting leak in the pipes under the house.

For years, you have observed a spreading wave of faint brownish or reddish-brown stains on your ceiling. At first, you may have thought that it was caused by a weeping nailhole in one of your shingles. That *is* a possibility, but as symptoms multiply, the awful truth begins to seep through at about the same rate as the staining on the ceiling. Your old steel pipe has had it.

Start your examination at the street, where the water comes in from the meter. If you're on a well, begin at the pump. Expose sections of pipe, especially in places where you've observed dampness in the past. Look for rusty pinholing, weeping joints, long lines of rust on vertical pipes, and "halos" of darker wood or stained sheetrock around the places where pipes enter walls or ceilings.

It won't take long to form your opinion. What may have started out a localized problem, with a bit of dripping from a ceiling, or a wet blossom on a wall, now falls into the context of an overall deterioration of your water system. Your old galvanized-steel pipes are shot.

Right then and there, get a pad of paper, pen, and tape measure and begin making a rough linesketch of your water-supply system.

There's only one way to do it and that's correctly:

Determine the size of your house main. If you have difficulty figuring out whether a pipe is a ½'', ¾'', 1'', 1¼'' or 1½'', the only thing you can do is measure its diameter, after you have scraped off a section to bare metal. With galvanized pipe, there could be some confusion, since the outside diameter is deceptive. For instance, the outside diameter of ½'' pipe is .84'' (a bit over ¾''). The outside diameter of ¾'' is 1.05'', of 1'' is 1.315'' (more than 1¼''), of 1¼'' is 1.66'' (over 1½'') and of 1½'' is 1.9'' (nearly 2'').

There are very few house mains plumbed in ½'' (it's too small), 1¼'', or 1½'' (they're too large). Most homes have ¾'' mains, and some very large places are provided with 1'' incoming lines.

You must replace all piping size-for-size. You may upgrade in size, but you should not downgrade. If it's 1'' now, your replacement should be 1''.

It is a fact that copper is rated at higher capacities than steel pipe, because the steel pipe was sized to take into consideration the corrosion which would diminish its capacity over the years of its life. Steel has a larger initial waterway than copper, but its surface is rougher, which affects its flow characteristics from the start.

The waterway of copper pipe holds its original diameter throughout its indefinite service life. Really, the only things that affect copper are the undissolved solids in water, like calcium, and other hard mineral substances which tend to build up in any pipe. But corrosion is not a major concern in the transmission of potable water through copper pipe.

By now, you should have at least found out the size of your house's water main. Next, start tracing the patterns of the water-supply network to the various sections of your home.

If it is a "slab-on-grade" structure, you may have problems doing that if the water lines were laid beneath the slab. In that case you may be faced with the total abandonment of the old system, and a whole different plumbing plan from the original.

If your house has an exceptionally limited crawl space beneath it, and you are a very large person, you may be faced with getting a short, skinny family member or relative to do the underhouse work.

If you can't get help from your family, you had better resign yourself to paying someone else to do portions of the job.

If your house has adequate crawl space, or most of the water supply comes from overhead, you should be able to make some sort of diagram, laying out the existing system.

As you are measuring the lengths of lines, noting their sizes, and drawing in the locations of risers (lines going up to fixtures) or drops (lines dropping down to fixtures), you should be thinking about demolition.

It has always been my policy to remove all old piping to be replaced. Not only does it make for a neater job, but it marks your work as "professional." However, sometimes, it is impossible to remove piping for one reason or other. I shall deal with that at the end of the assessment portion of this phase of the work.

Speaking of demolition, you must also be noting places where the removal of walls, partitions, ceilings, and other structural members will be mandated by the installation of the pipe services.

Unless you are extremely skilled in construction, you should have the structural, drywall, and finish work seen to by experienced people, whether they be professionals or skilled friends whom you can entice to give their all on the promise of a keg of beer and a barbecue, *after the job is done!* Too much beer during construction has resulted in more bulging walls (not to speak of bellies) and sloping ceilings than any other problem I can imagine.

The important considerations during this phase are the easiest, simplest, and most direct methods and pipe-laying routes required to achieve the desired results.

Realize that, if you are able to remove the existing pipe, you will be able to follow the original routes exactly, which should simplify things considerably.

Most houses have ¾" submains and ½" services. I, personally, do not run ½" to two fixtures, but I have built very long ½" runs to feed a single fixture.

There are rules of thumb for sizing pipes, and they are all based upon consumption. Just think about it—Which fixtures in your house use the most water, and which use the least? High-consumption units are the tub filler, kitchen sink (especially if it has a dishwasher), the laundry complex, with its washing machine and tubs, and all hose bibbs.

Showers, lavatories (bathroom sinks), tank-type water closets (toilets), and accessory sinks (like wet-bar sinks and bidets) are low-consumption units.

Each high-consumption fixture must have its own ½" hot-and-cold water supplies. In a pinch, you could connect two low-consumption fixtures to the same ½" services, *but no more than two.*

CAUTION: *When I suggest that you may hook up two low-water consumption units to the same ½" line, I presume that you have decent water-supply pressure (over forty pounds per square inch). If your main pressure is lower, stay with one fixture per ½" service.*

Picture this: You are coming into your house with a ¾" main service from the meter. You continue that ¾" size all the way to the water heater. In the course of running that line, you pass places which have service risers or drops; you take off with a ½" pipe for each service. If you pass a point where two fixtures are in line, one close to your main, and the other farther away from it, you should take off with a ¾" lateral from the ¾" submain. After you have taken off your first ½" service, you can then go to the next fixture with ½".

The reason I'm going through these details now is because I'm aware of the fact that a lot of old-time plumbers and the great majority of home craftspersons have plumbed tandem and triple steel services in ½". That practice has resulted in "starved" fixtures when two or three of them, on the same ½" line, have been used at the same time.

If you were working in galvanized-steel pipe, you'd know why they did it. Half-inch pipe is lighter, easier to cut and thread, and a great deal easier to fit up than ¾" pipe. It is also less expensive.

In the beginning, most people wouldn't have any problems. Then corrosion begins, and the walls thicken with oxides and deposits. In a few years, the decline in flow becomes noticeable. In fifteen years, the system shows all the symptoms of arteriosclerosis. The waterways have closed down to the point where drawing water from one fixture will deprive another one on the same line.

To have a well-designed system, simply keep the interior submain size the same as the main between the meter and the house.

HELPFUL HINTS: There are always exceptions to rules. One of them would be in the case of an incoming main of 1" or larger. Such a line implies a house of considerable size, certainly larger than the average three-bedroom, two-bath tract house; however, even that observation could be erroneous. Home craftspersons, conscious of the value of installing larger pipe sizes to increase the flow-patterns in their homes, have been turning to larger pipe to remedy problems relating to "fixture starvation." It's a good idea. If you're doing the work yourself, the only increase in cost is the material. The labor cost is the same: nothing.

Think about running a 1" main from the meter.

If you enter with 1", you can split into two ¾" submains, which could make your service runs shorter, neater and more efficient.

Since we are discussing copper pipe installations here, I include the outdoor main because many of you will prefer to install a copper house main.

I had suggested earlier that the main between the meter and the house could be PVC plastic pipe. That is still a good idea. But, when it comes right down to push-and-shove, copper is by far the better material.

By this time, you have probably thought through the problems to which I've alerted you, and you're ready to go forward into the next phase.

To recap: You've decided to construct a new water system for your house. Considering your circumstances, you've fixed in your mind whether or not you plan to finance the project. You have located your house main (the line between the meter and the street), and determined its size. You have followed that line into the house and traced it, noting the locations of services. Throughout the process, you have made notes and sketches, indicating the location and function of each lateral, the pipe sizes, and valve placements. You have made a rudimentary plot-plan, showing the size and shape of your lot or property, and then the shape and dimensions of your house and outbuildings. It is best to do these drawings to some sort of convenient scale, a selected fraction of an inch to each foot of actual dimension. In that way, the size relationship of the house to the land will be shown accurately.

HELPFUL HINT: You don't need a drafting table or other exotic mechanical-drawing equipment. An ordinary ruler and a piece of 8½" x 11" typing paper will do very well. If you want to make your drawing to a larger scale, or if your property is quite large, just get a larger piece of white paper or reduce the scale.

The important thing is getting the whole project down on paper in a form that will make the presenting of your proposal to your lender, the obtaining of a permit (if you decide to take one), the purchasing of the materials, and the building of the job easier and more professional.

MATERIALS LIST

Before you can make a materials list, you must consult the local plumbing code. Even if you don't take a permit, you must *follow the code strictly. If you ever want to sell your house, and you are in an area which requires a building department compliance certificate prior to sale, for the protection of the buyer, plumbing to the code can mean the difference between an uncomplicated routine sale and a costly revamping of the system to bring it up to code. You may save a few dollars at the time you build, but in ten years or so, even with higher materials costs, those extra forty pounds in the hips and tummy can send you to a pro, who may just sock it to you right where you live, knowing that you have to do the work before you can sell!*

The information I am about to lay on you comes from the *UPC*, which should cover your community. If it doesn't, write to National Plumbing Information Service, Post Office Box 3455, San Rafael, CA 94912-3455. They maintain a current file on the codes in force in the United States and Canada. You'll probably have to wait a couple of weeks for their response, so keep that in mind when you set up your timetable. Any inquiry of the service has a straight charge of $1, except for publications and books, which they will quote to you if you send a self-addressed, stamped envelope with your request.

You can avoid having to spend the buck, if you'll take the time to phone around City Hall or the County Court House and ask a few simple questions.

The *UPC* permits the use of type M copper water tube (Sec. 203), but its use is restricted to *above ground inside of a house, or underground outside of a house.* The same would apply to your outbuildings. CAUTION: *Type M is not permitted in commercial or industrial buildings in some jurisdictions, even though it may be permitted by the* UPC.

In any event, you are always safe with type L, which has thicker walls.

You are required to use type L whenever your pipe is exposed on the outside of the building, and when you bury it in the ground beneath your house. If you pour a slab over copper plumbing with the intention of building over that slab later, it's best to be safe and use type L under the concrete.

Copper water tube of all types can be purchased either as a rigid pipe or as soft (annealed) tubing. The annealed tubing comes in coils from 50 ft. to 100 ft. It makes running the house main from the meter relatively easy, since there are no joints between the two points. All you need do is make the connection to the meter and run the line to the house without connections, in most cases. If your home is 200 ft. from the meter, we're talking about *one coupling*. Considering how many fittings you're going to have to install inside your house before you're done, this job is a piece of cake.

From your drawings, measure off the amount of pipe you'll need. Separate it into the two categories: (1) Pipe in the house, not buried in the ground. Pipe buried in the ground outside. (2) Pipe exposed on the outside of the building. Pipe buried in the ground inside the house.

Category 1 needs type M copper water tube. If you wish to buy the soft tubing in coils for the outside house main, it comes in type L and M. Otherwise, buy normal, rigid type M copper, in 20-ft. lengths. Category 2 needs type L rigid copper water tube in 20-ft. lengths.

Now for your sizes. In most cases, your house main will be ¾″ or 1″. The submain in the house will be ¾″ as well. The services to fixtures and bibbs will probably be ½″. All of this pipe should be rigid, and in 20-ft. lengths. There are times, however, when the soft tubing comes in handy. For instance, when you have a long run from one part of the house to another without too many fittings, because of the configuration of the space through which it runs, a straight line would not do the best job. With plenty of twisting and turning the annealed tubing is the best material to use. However, if there are a lot of fittings to be soldered into the line from point A to point B, stick with the rigid tubing. The soft material has a tendency to deform on the ends when it is cut with the tubing cutter, which makes a difficult fit-up. There are ways to true-up (make the ends perfectly round once again) "end ovals," but they take time and experience.

Left, top: 90-degree copper elbow; *right, top:* 45-degree copper elbow; *left, center:* 90-degree copper street-elbow; *right, center:* 45-degree copper street-elbow; *bottom:* copper tee.

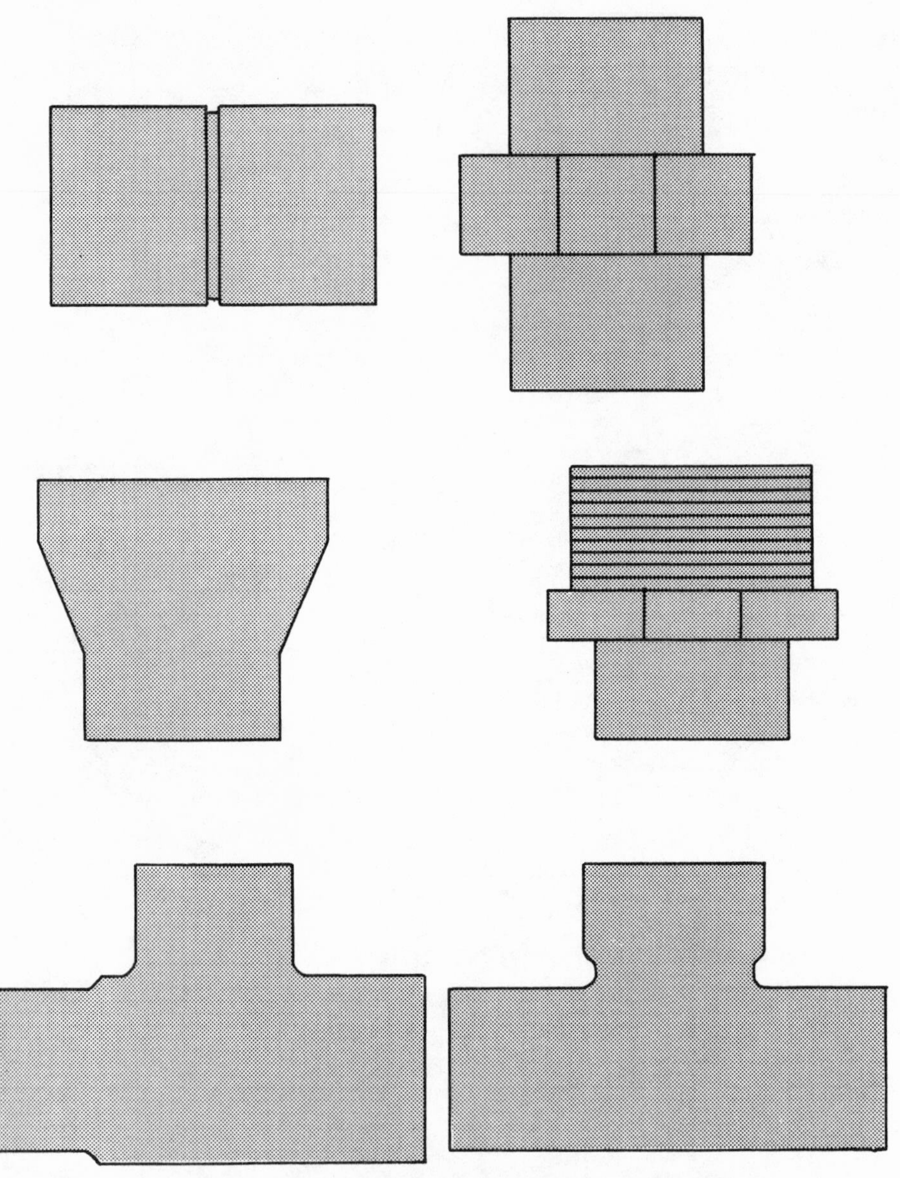

Left, top: Copper coupling; *right, top:* Copper union; *left, center:* Copper female adapter; *right, center:* Copper male adapter; *Left, bottom:* Copper reducing tee (reduction on run and outlet); *right, bottom:* Copper female-threaded-outlet tee.

You have your types and sizes taken off, and now all you have to do is consolidate them into two or three quantities for purchasing purposes.

Let's talk about the fittings.

The normal fittings for copper residential plumbing are the 90-degree elbow, the 45-degree elbow, the 90-degree street elbow, the 45-degree street elbow, the tee, the male adapter, the female adapter, the cap, the tee with female-threaded (FIPT) outlet, the 90-degree elbow with female-threaded (FIPT) outlet, the dielectric union, the ground-joint bronze union, and the coupling.

These are made in straight 1″, ¾″, and ½″ sizes, and in reducing sizes. An example of a reducing tee would be a ¾″ x ¾″ x ½″ tee. The first two numbers refer to the run size, which is the part that goes with the run of the original pipe. The third number is the outlet size, to which you attach the branch that you are connecting to the run of the pipe. Another example would be a 1″ x ¾″ x ½″ tee. This is a very useful type of fitting. Wherever it occurs you know that the main line, or run of the pipe, is changing size. It comes into the tee as 1″ and goes out as ¾″, with a branch outlet of ½″. A bull-head tee is one which has a larger outlet than its run. Such a tee would be specified as ¾″ x ¾″ x 1″. The run of the pipe is ¾″ and the outlet is 1″. You use a tee like this when you are bringing in a main line and splitting it into two submains, and in other similar circumstances. When you are asking for a copper soldering tee with a threaded outlet, you ask for a ¾″c x ¾″c x ½″ FIPT tee. That translates: ¾″ copper by ¾″ copper by ½″ female iron-pipe thread. The two sizes of a reducing elbow are always listed, larger size first.

A reducing female adapter would be called out as a ¾″c x ½″ FIPT, which translates to ¾″ copper by ½″ female iron-pipe thread. Or you could have the FIPT larger and the copper socket smaller. A reducing male adapter would be ¾″ MIPT by ½″c. There are also the common or garden variety ¾″ male and female adapters.

The object of this exercise is to get you speaking the plumber's language. If you are able to keep your terms straight, and write them correctly, you should be able to order exactly what you want. That is more important than you know.

So many home craftspeople just don't come off knowledgeable when they go to a store to buy material, and they are "branded" by the people who help them as putterers and tinkerers. You'll be appreciated by your local supplier if you come into his place and can state your order in clear, concise, technical language. Get used to calling fittings by their proper designations. Watch those clerks show respect as they put up your order.

Your order is ready for the pipe and fittings. It is time to get on with the rest of the goodies. A half-box each of ¾″ and ½″ copper pipeclamps, with some copper brads (nails) for the clamps, plus a 25-ft. roll of copper plumber's tape would be your next purchase. Naturally, if you were going to install some 1″ pipe, a dozen or so 1″ clamps would come in handy.

I expect that you've bought the solder, steel wool, and flux recommended in Section Two. CAUTION: *Buy only the non-corrosive flux required by the code. Corrosive flux can cause leaks years later.* While you're at it, it would be a good idea to buy about a dozen small acid brushes. They are used to apply the flux to the pipe. These brushes will save you a lot of grief in the application of flux to polished pipe ends. They cost about fifteen cents apiece and are worth every cent of it. I usually throw one away after each day's work. They are not reuseable with most types of fluxes. They accumulate old, clumped paste-flux residues, and have a tendency to break down the new flux, after a few days of use.

Naturally, you have your tools, including your tubing cutter, torch head, propane or Mapp gas, for whatever type of head you've bought, measuring tape, and a green felt-tip pen.

I haven't mentioned the pen in my tool list because, in fact, you can use just about any kind of felt-tip pen. The green medium-tip seems to work best for some reason. It goes on copper well and makes a great markerline.

It would be well to have your hacksaw handy as well.

That's it for your "rough" plumbing. When you get to "finish" plumbing (hooking up fixtures), you'll need a few more items.

Before you go downtown to see your lender, you should have your purchase list ready.

On all pipe, fittings and soldering supplies, figure actual cost, tax, mark the whole thing up around 15 percent, and that should do it. The markup is what we call "cover-your-rear money," to keep you on the plus side, if you forget an item or two.

A good example of forgetting an item or two has just occurred. I forgot the valves and hose bibbs, which are calculated in the "rough" plumbing costs.

You will need a valve the size of the house main, to be installed where it comes into the building, on the outside. That should be a gate valve (see the illustration and description of the gate valve in Section Three).

There must be a gate valve on the cold-water supply side of the hot-water heater. As I explain in Section Three, it's a good idea to put a gate valve on the hot-water side as well.

Count the locations of your hose bibbs and make sure you buy enough of them, plus one. It has been my experience that, just before you wind up a job, it is inevitable that someone is going to come up with some off-the-wall additions, usually hose bibbs. The usual reason given for the addition is, "Honey, it sure will come in handy, when . . . " Take my word for it, get an extra hose bibb.

If you have a clotheswasher, you'd better get a pair of special washing-machine valves, and last but not least, buy two globe valves (see Section Three for the illustration and description) for each shower, and bathtub, if they are separate. Most plumbers don't install shutoff valves for showers and tubs, but I do and I suggest that you do too.

If you're going to finance your job, you probably have no choice in the matter of obtaining a permit. Most lenders just simply won't grant loans unless the jobs they cover are officially regulated. It's a protection for both the bank and the client.

I've already spelled the procedure out in Section Four, and I've reinforced my remarks in this section. I suggest that you reread the relevant passages, and get cracking.

Once you have your permit and financing bases covered, you're ready to proceed with the actual work.

DEMOLITION

To remove the pipe, or not to remove the pipe, that is the question.

The only times I recommend against removing the old galvanized-steel pipe are when it is physically impossible, or the risk is excessive.

It is impossible when it's buried in a slab. It is difficult to the point of being too risky when it runs through structurally sealed joists, between the floor above and the ceiling below.

The only other valid reason for not removing the old material that

I can possibly conceive, is the pressure of time. If you have a crisis, and you must get on with the job as quickly as possible, I can understand leaving the pipe in the walls, and then running the new pipe parallel. But, there is no excuse for not removing all the visible pipe.

Disconnected steel pipe hanging about can be a hazard. It has no use whatsoever, and can obstruct the progress of your new project.

HELPFUL HINTS: At this point, it is important to make arrangements for the continued daily operation of your home, if you plan to live in it during the period of the plumbing project.

For sanitary facilities, I suggest an outdoor temporary chemical toilet, of the type used by contractors. They are listed under *Toilets—Portable,* in the yellow pages.

Keep your house system in operation as you lay your house main parallel to active portions of the existing steel line. When you have the line in place, fit it up and make it ready to connect at both ends and shut everything down for the period of time it takes to put the new main in service.

You are going to have to accomplish your demolition in one weekend. Some assistance is necessary, if only to hold, fetch, carry, and help you measure and cut.

You have water to your house through your new line, or, if you haven't abandoned the old main, the old service remains active. If your main doesn't have a gate valve on the outside of the house, put one there. Just before the valve you should have a hose bibb or faucet of some sort. That faucet is going to be very important to you during the time of your project.

From that bibb you can take a hose line to supply your household with drinking and washing water, while the job is in progress.

Removal of the steel plumbing in the house can be accomplished with a veritable arsenal of tools. The fastest and easiest is an acetylene cutting torch. It takes some expertise, and a bit of practice, but it is fast. CAUTION: *Always keep a fire extinguisher, bucket of sand, and/or container of water handy when using hot-firing torches of any kind, and that includes your propane or Mapp gas soldering torch.*

If you have no experience with a cutting torch, don't even think of trying to use it.

The next most obvious option is to cut the pipe out with some sort of a saw. The following types of power saws are highly recommended: Milwaukee Sawzall, Black and Decker Cutsaw, and other similar types of reciprocating saws which you can equip with 32-tooth metal-cutting blades. SPECIAL COMMENT: Blades coarser than 24-tooth are difficult to use, and tend to lose teeth rapidly. You can use blades as coarse as 18-tooth, but you'll be in for a "rough ride," with frequent blade replacements.

If you can find one, the Milwaukee portable band saw is the Rolls Royce of portable saws. Its action is smooth and steady, and it makes mincemeat out of pipe up to 3". However, it is difficult to maneuver in very tight places, and it is not readily available from rental yards.

Again, try to get close to 32-teeth per inch in your steel-cutting blade selections. With power saws, a 24-tooth blade is acceptable. As I said before, 18-tooth blades are marginal.

And then there is the hacksaw, operated by person-power. It's reliable. It gets the job done. And it gets into tight places.

If you're a lawyer during the week, and the only work you do is sedentary, manually hacksawing the steel pipe out of your house will either put you in the hospital, or turn you into one of the fittest persons in your set. You'll acquire sore muscles you didn't even know you had.

For very tight places, there is a special handle available from the hardware store. It literally clamps onto the blade, eliminating the need for the normal hacksaw frame. *Use this device only when required in very close spaces.* The handle doesn't give the blade as much support as a conventional frame.

HELPFUL HINT: It is completely unnecessary to cut anything out with a view to possible salvage. *Don't even think of that!* Cut the material out in as large sections as you can. Where possible, turn sections out with your largest pipe wrench, equipped with a "cheater," if necessary.

In demolition, safety is first and speed is second. All other considerations are farther down the priority list.

CAUTION: *Your biggest danger will be from falling objects, not necessarily the pipe itself. It is positively amazing how many things—chunks of debris, old skis, boards, loose pipes, and other miscellaneous junk—are held in place by water pipes that have been called into service as overhead racks. The original idea was that the arrangement was temporary. The years pass, and that temporary arrangement has assumed a permanency that only forgetfulness can insure.*

FURTHER URGENT CAUTIONS: *When you work at demolition, wear safety glasses, gauntleted gloves, and substantial clothing and shoes, preferably safety shoes. A hard hat is strongly recommended where there is the possibility of heavy falling objects, or when another person is working with tools above you.*

Have a first-aid kit handy, and I don't mean just a small box of bandaids.

Take a ten-minute break every hour, and if you're working in crawl spaces full of dirt, dust and debris, try to get out into the open air, especially if you are using torches.

In black-widow country, be on the alert for this deadly spider. If you know that you have a spider problem, it may be best to fumigate before you undertake your project. If you decide not to do that, then by all means carry an aerosol of appropriate control with you.

In very dusty places, wear a respirator. When spraying for black widows, use a respirator designed to keep the spray out of your lungs.

The same cautions apply for bees, hornets, wasps, and other stinging insects and animals.

The other day, an acquaintance came to me for advice on her renovation job. I gave her the same hints and cautions I am giving you right now. She laughed at me, and spent a considerable time telling me that caution was her middle name. She was at pains to convince me that she was among the most safety-conscious of her friends and co-workers.

In cutting out a length of gasline with my Cutsaw, she managed to cut a support simultaneously, which shook loose a wasp or hornet's nest just outside a window.

She's just come out of the hospital after a touch-and-go bout with anaphylactic shock.

Six or seven years ago, one of my best journeyman plumbers was bitten by a black widow. He almost went down for the count.

One of your biggest frustrations in demolition will come when you try to remove a line that has been passed through drilled holes in joists. Looking at such lines, at first you can't understand how the plumber got them in place in such long runs.

If you cut both ends loose on such a run, and try to pull it out, you find that it's impossible. You can't get the pipe out in one, two, or even three pieces. The joists are holding the entire line fast, at 12″ to 24″ intervals (usually 16″ or 20″).

Obviously, the installing plumber set his pipe when the house was nothing but open framing, so he could drill the joists straight across from outside the building, using very long augers to start, and then taking one or two studs at a time with his other bits.

HELPFUL HINT: There are rules of thumb for drilling or notching supporting timbers and other structural members. *Drilling:* The diameters of your holes may not exceed 40 percent of the width of the board being drilled, if it is a *non-bearing member.* The limit is 25 percent *for bearing members. Notching:* The notch *may not exceed a depth of 25 percent* of the width of the board being notched, if it is a non-bearing member. It is 15 percent *for bearing members.* Every building code, building department, and town or county engineer's office has a more specific view on this.

CAUTION: *Before you drill or notch any structural boards in your house, make it your business to find out what the local limits are, and don't exceed them. Once you've cut or drilled, the only way to repair a mistake is costly and time-consuming.*

SPECIAL COMMENT: A bearing member is a timber, joist, stud, sill, or plate that bears the weight of the house itself. Some people take the view that a cut here and a hole there doesn't really mean all that much, considering how much structure remains to support the building. That's a crock!

Amateurs who take that view are exposing themselves to the most horrendous experience imaginable, the profound sagging or even collapse of their homes.

I have been called an alarmist by some, but I've seen it happen.

I can't impress upon you enough the value of observation. Whenever you are exposed to the skeleton of your living place, *look at it! Familiarize yourself with your environment.*

As you're going about the business of demolition, take special note of the amount of notching and drilling that has gone on over the years. Do a little mental arithmetic, make assessments and store them in the front of your consciousness, while you work close to the vitals of your home.

HELPFUL HINT: When you replace the pipe in your house, try to arrange your pipe runs in such a way as to avoid cutting and notching as much as possible.

Copper pipe is considerably lower-profile than steel, size for size. Take advantage of that quality.

Now, let's get back to the pipe, laid through a succession of joists set at 20″ intervals. If you have a cutting torch or power saw, just go right down the line and cut the pipe between each joist. You'd be surprised how fast the whole thing goes.

If you have to do it by hand hacksaw, that's a horse of another color.

Perhaps you should speak to friends about cutting in shifts. It will be a long day, believe me.

PIPE REMOVAL FROM WALLS AND CLOSED SPACES

This part of the demolition process is the most difficult and scary. It implies cosmetic scarring and open, dirty spaces that have been long-concealed and out of mind.

There is always fear of removing something that you may not be able to restore to its former condition.

Right here and now I'm going to tell you not to worry. If you can't restore it properly, then restore it the best you can, and hire someone else to put on the finishing touches. It won't cost that much!

The matter of paramount importance here is that you approach the work *without fear!* If you don't, you'll never finish the job. As a matter of fact, if you can't control your worry and fear of making an irreparable mess, don't even start the job.

You can minimize the mess and damage, if you clear out all of the rooms that will be affected by the demolition process: the kitchen, bathrooms, laundry, utility rooms, and, if you have one, the wet bar.

Buy a bunch of disposable plastic-coated paper or all-plastic painter's dropcloths. Secure them to the baseboards, if necessary, with masking tape.

Before you break out a section of wall or ceiling, survey the situation carefully. Is it possible to feed up the pipe, through the holes in existence? You will determine that from the ease with which you removed the old pipes to begin with.

Say you've disconnected a ½″ cold-water service which goes to your kitchen sink, from below, where it joined the main. You have unscrewed the angle stop up on top, and unscrewed the nipple to which it was joined. Now, all that's remaining in the wall is the pipe, the elbow which held the nipple to which the angle stop was attached, and any couplings that may have been used by the installing plumber to extend his service.

When you look up from down below, you just see a hole in the sill or, perhaps, in the space between studs. If the service goes through the sill, plate, or other structural support underneath the kitchen, it's best not to enlarge the hole and attempt to fish the pipe out from below. If, on the other hand, the pipe comes down between studs, and doesn't appear to have been put through any bearing member, you can enlarge the hole to as much as 2″ in diameter, and try to fish the riser through.

CAUTION: *Quite often a wall service riser can extend above the first fixture it serves. Remember, I mentioned that many of the old-timers put two and three fixtures on one ½″ line. Most often they did it vertically rather than laterally. Also, you should know that, for years, plumbers were required by various codes to install "air chambers" on all service risers and the water heater. "Air chambers" were made onto the run of a tee. The outlet of the tee held the nipple to which the fixture's angle stop was fastened.*

An "air chamber" is just a 12″ nipple with a cap on one end, screwed into the tee at the top. The service riser from below comes into the bottom of the tee. The outlet holds the service itself.

The problem with these "air chambers" is that many of the old-timers used the "air chamber" nipple as a place to secure the service with a clamp. The clamp was placed between the tee and the cap. There's no way you are going to be able to fish the old service out from the bottom.

The same applies to services that are continued up to the next floor to supply other fixtures. When you're up against this problem, it's best to go in and open up the wall, and be done with it. The same may be said for any extensions from or to any riser.

In some installations, the submain runs in the walls. When that happens, I seriously consider abandoning the removal of all concealed pipe.

As soon as removing the steel pipe becomes a tremendously time-consuming chore, I recommend that you remove what you can see, and reroute the new risers and other submains.

Opening up lath-and-plaster or solid-wood-sheeted walls is the most troublesome chore. Later, these kinds of walls are the most difficult to refinish to exactly the same texture and appearance.

In front, I suggest that you consider surfacing the wall in sheetrock (drywall), rather than trying to match the existing finish.

You should be able to lay the sheetrock right on top of the present surface and your pipe-exposing cuts. It's a very professional method of inexpensive repair to difficult-to-refinish walls.

If the wall is sheathed in wood, veneer-type plywood or composition board, perhaps you can remove the sheets, intact, and cut beneath to expose the pipe.

In this case, you don't have to worry about closing the cut. Just fill the space with insulation, and replace the sheet. That method makes a fine solid repair.

In the kitchen, much of what goes on happens in the backs of cabinets. It's obvious that covering up down there is simply a matter of cutting panels to fit and drilling holes where required.

When you cut sheetrock there is a special procedure to be observed. You must cut to the studs so that you can put in a solid patch when you're through with the plumbing.

By cutting to the studs, I mean that you should locate the studs, either with one of those inexpensive little plastic "stud finders," or you should measure from the corner of the room, and calculate in multiples of 16", which is the standard centerline distance between studs, in most codes.

Often, simple observation will provide the clue. Have you ever looked closely at your walls? If you have, then you may have noticed one of two things. Either there is a tape outline or a line of nail pocks evidencing a stud or the joint between two pieces of sheetrock. The crack where two pieces meet should lie right over the centerline of a stud. The nails on the ends of the butted sheets should lie on both sides of that centerline.

Now, if you calculate 16" intervals from one of those centerlines, you should hit a stud at every interval. A single line of nails, one on top of the other, indicates a nailing on the stud. Of course, older houses could have studs set at different intervals.

Cut as close to the centerline of a stud as is humanly possible.

Cut a 16" piece out in as perfect a rectangle as you can manage. If necessary, take the cut from floor to ceiling, cutting on a straight-edged implement, like a metal yard-stick, if you have one.

If you can take the piece out cleanly, perhaps you can save it as the patch.

In any event, the object is to leave a nice clean ledge of stud upon which to seat your patch, for a sharp-edged, precise joint.

Because of a peculiar kind of psychological inhibition, people want to cut the smallest holes they can in walls, thinking that they will be less noticeable when patched. *Wrong!* The best patch is a nice, smooth transition from one surface to the next. The stud-to-stud cut is *by far* the most professional technique.

Think about it; if you cut too small a rectangle out *between the studs, you'll have to fasten to blocks.*

Also, as in medical surgery, don't be afraid to "get it all." Working in a restricted space can often frustrate operations. When working in walls, you must have wrench and torch room. Without it, the work can become an exercise in futility, shortening tempers and causing mistakes which can be costly later.

I used to use the term "rip-and-tear" when I referred to piping demolition work, and people would shudder at the thought. Most folks think of their nice, neat homes as the rocks of their existence, immutable anchorages where change is to be avoided, if at all possible.

Putting a 1-ft. x 9-ft. gash in a paneled or freshly painted wall, and exposing all sorts of smelly, musty crap, which is normal for a house that's been around for awhile, is more than some people can bear. Yet, bear it they must, if they're going to avoid much greater and more dangerous messes later.

Get on with it! Pull those pipes out of the walls.

CAUTION: *Do not leave dismantled lines anywhere that can cause a danger to people walking on the property. Don't let that developing scrap heap of yours become what is called an "attractive nuisance." If some kids start to play in it and they are injured, you will pay through the nose. That is grounds for suit and I've seen some pretty outrageous judgments handed down in such cases.*

Frequent trips to the dumps are in order.

When you remove the pipe from the various fixtures, faucets, valves, and bibbs, don't forget to strip your water heater. Take everything off right down to the spuds (nipples coming out of the heater). Examine everything carefully, and discard anything that seems to be worn badly.

INSTALLING COPPER PIPE

Before you cut your first piece of pipe or burnish the first fitting to be installed, you must learn how to solder properly.

Prepare yourself. Screw your torch head onto a tank of gas. Get a pad of steel wool out of the pack. Open your container of flux and set an acid brush in the paste. Make sure you have a "sparker" or "striker," to light your torch. If you don't, they cost around $3 at the local hardware store. With this, you are able to light your torch without creating an open flame.

Have your tubing cutter handy, and all your materials laid out.

First, measure a 12" length of ½" pipe, and mark it with your felt-tip pen.

Set the cutter wheel on the mark and start tightening the screw, bringing the wheel into biting contact with the pipe. As you do that, begin to roll the cutter around the pipe, making a shallow groove in its surface with the cutter wheel.

As you turn the cutter around the pipe, take up on the screw handle just a wee bit each revolution, so that you increase the groove in the pipe surface. After awhile, you will feel the tube beginning to give. Finally, it will part, just like that. You've made your first cut.

Cut another piece of 6" in length, and a third of 4".

Now, get one of your ½" tees. Set down all four pieces in front of you, the three bits of pipe and your tee.

Take the steel wool and gather one corner into a cone by twisting the strands, and insert that corner into the outlet of the tee. Holding the tee firmly, turn the cone of steel wool inside the socket of the tee's outlet several times. Upon your removal of the steel wool, the tee's outlet cup should be brightly burnished. Do the same with the two run cups.

CAUTION: *It is important that you use the reamer on your cutter to eliminate burrs on the pipe ends.*

Next, polish the ends of the three pipes, by holding the steel wool in one hand, the pipe in the other. Now, simply lay the pipe end in the wool. Close the wool around it and revolve the pipe several times, until it is brightly burnished. When it is placed in the cup of the tee, both meeting surfaces will have been properly prepared.

COPPER PIPE SOLDERING PROCEDURE Top illustration, this page: *Step 1—Using medium steel wool, polish the outside of the pipe end to a bright copper shine;* bottom: *Step 2—Polish the insides of all fitting outlets until they're bright.*

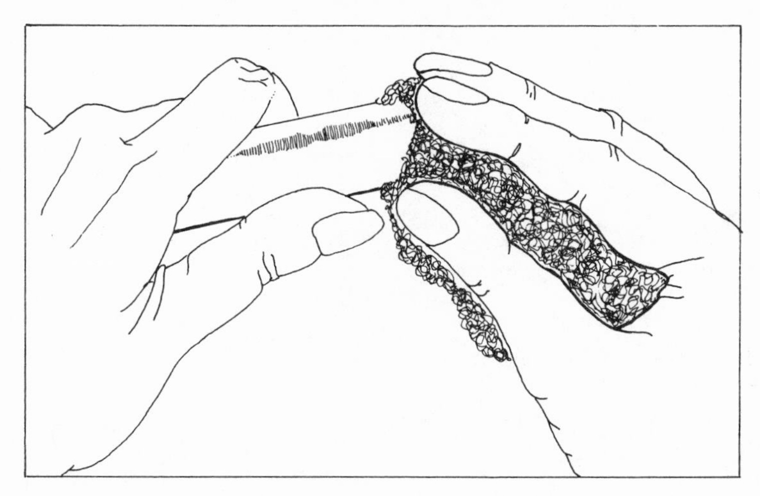

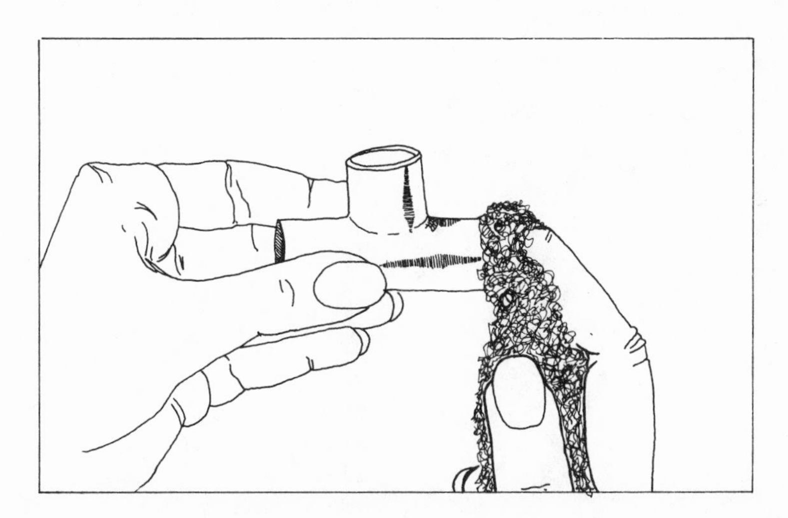

Two hands holding copper pipe and steel wool, polishing the outside of the pipe in the top frame, and polishing the inside of the fitting in the bottom one.

Top illustration, this page: *Step 3—Brush flux evenly on the outside end of the pipe;* bottom: *Step 4—Flux the insides of all pipe-fitting outlets (not required on fittings under 1¼"; Step 3 is sufficient).*

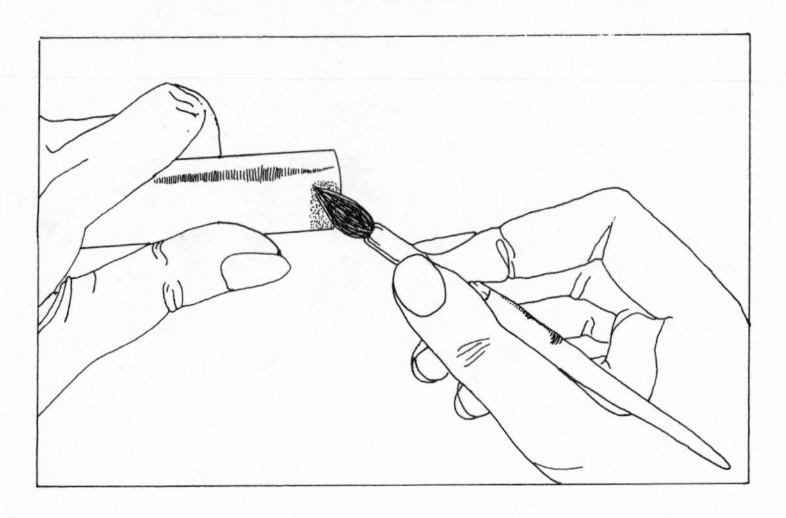

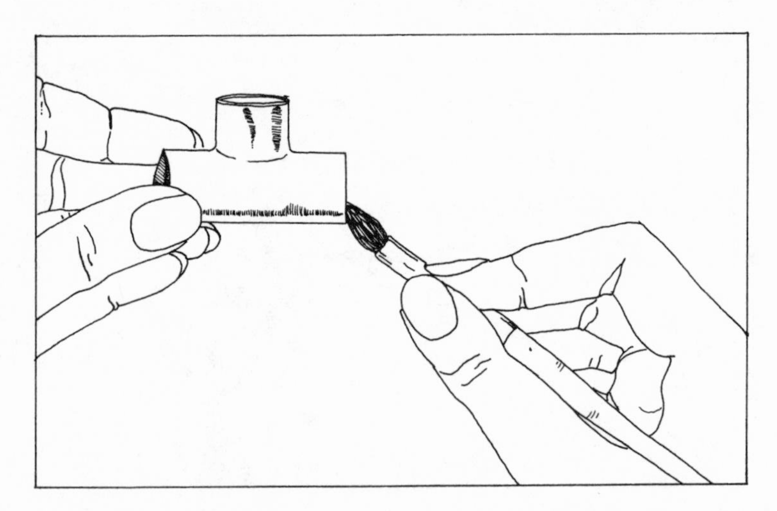

Two hands holding copper pipe and flux brush, applying flux on the pipe in the top frame, and into the fitting in the bottom one.

Top illustration, this page: *Step 5—Apply propane torch heat on the fitting only, well back from the outlet being joined, and when the flux bubbles and dries, silvers, or begins to smoke, go to Step 6;* bottom: *Step 6—Apply solder to the place where the pipe and fitting join, working the solder around quickly to fill the fitting cup (socket). When the first drop of solder falls out of the cup, it's finished.*

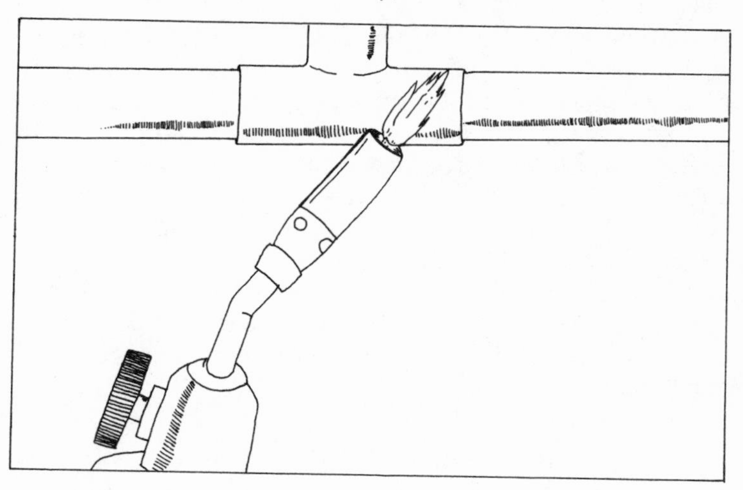

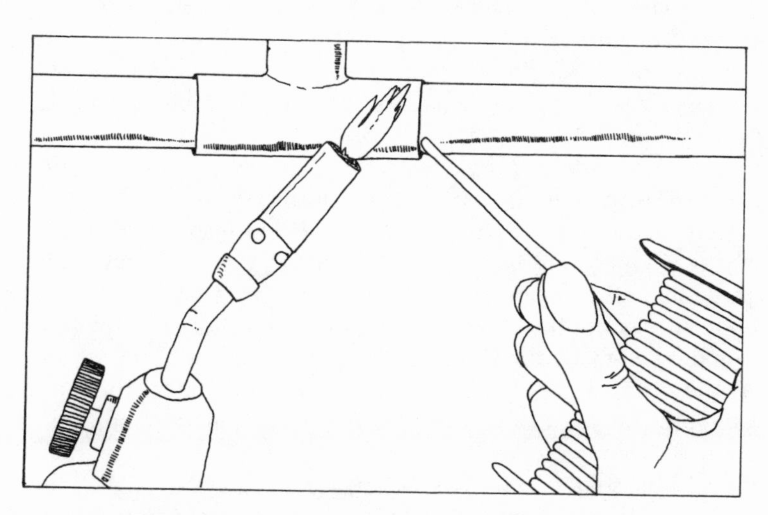

Top: Flaming torch being directed to copper pipe-fitting; *bottom:* Hand holding solder spool, and flaming torch.

At this point, you coat the pipe ends with a thin film of the flux, where they've been polished.

Place each piece of pipe in one of the tee's cups.

SPECIAL COMMENT: You should place the 12" length of tube in a vise, or brace it so that the piece of pipe is down, and secured. It would be a good idea to have the 6" nipple up, and the 4" one coming out of the tee's outlet. The object of the exercise is to teach you to solder at the three most common positions, vertical-top, vertical-bottom, and side or horizontal.

SPECIAL COMMENT: Now I'd like to tell you about a personal technique that I have come to feel is the best I've yet discovered. It has always worked very well for me, and I hope that it will for you. (That doesn't mean that there aren't other techniques, some of which may be every bit as effective.)

In time you will develop your own skills and techniques, but starting with mine, you may be able to develop your own successful variations more quickly than I believe possible with other effective methods.

SOLDERING TECHNIQUES

The phenomenon which causes the solder to be drawn into the fitting is called capillary attraction.

The fittings are a predetermined number of thousandths of an inch larger than the pipe, which is a factor in properly made and joined fittings.

The cardinal rule in tubular soldering is always to heat the fitting and not the pipe. Repeat: Don't heat the pipe!

Of course, it is inevitable that the pipe will heat, when you are applying over a thousand degrees of heat to a copper fitting, in intimate contact. But the pipe mustn't get as hot as fast as the fitting, or the solder may not be drawn in to fill the entire cup as completely as it would if you made the fitting much hotter than the pipe.

As the fitting heats, it expands a bit, and literally sucks the molten solder into the socket. So strong is the attraction (suction), that it even works when you are soldering a fitting from below. The hot fitting sucks the solder up into its socket, defying gravity.

The main thing to practice is the rapid heating of the fitting, and the proper application of the correct amount of solder.

Let's get to it. Turn on your torch, open the valve all the way. It is not a good practice to try to throttle the flame. Often, the valved-down flame begins to burn down in the tip instead of right at the tip orifice, resulting in a glowing torch-tip. A proper flame, which originates at the very tip, doesn't heat the tube to a glow.

When you open the valve all the way, you just hold the flame on the fitting for a shorter length of time. Or you apply the flame at a greater distance, to control the fitting temperature.

If your torch has a big, bushy flame that slops all over both the pipe and fitting, you'd better trade in the tip for one that puts out a pencil-width flame.

The most difficult will be your first joint: the bottom socket of the tee, in which you have inserted your 12″ tube.

The ½″ tee is so small that it's best to direct the flame on the center of the fitting. That will heat all three sockets at the same time.

You will notice that the flux will sizzle, and if you are using the recommended AMCO-C flux, it will start to get silver. As soon as the sizzle begins to lessen, and, certainly, when the silver appears around the cup, it should be safe to touch your wire solder to the crack made by the joining of the tube and the socket. If the solder doesn't begin to melt immediately, the fitting isn't quite hot enough. Continue to touch the solder to the joint, testing, until it begins to run freely.

Now, take the flame away from the fitting for a moment. The solder should continue to run, and you will notice that it is being sucked into the fitting. When you are soldering a bottom socket, as soon as the solder begins to run down the pipe in a beginning stream, the joint should be okay.

The other two joints are duck soup. First solder the top, and then finish up with the horizontal pipe.

The top one should take instantly. You will wonder why it seems that the top socket doesn't fill up with as much solder as the bottom one. There are several reasons for that, but suffice to say that it is normal. The solder will probably build up to a nice bevel.

Don't be alarmed if more solder begins to run down the pipe from the bottom socket. Some of that may be coming from the top joint that you've just finished.

As for the horizontal socket, apply it at or near the top of the socket. The moment the *first* drop falls, the joint is properly made.

At this point, it is best to wipe with a rag the three joints that you've just made, to remove the excess solder, and give the job a professional touch. Water should be applied to the hot fitting, in one of two ways. Either you can spray it from a plastic spray bottle, set to fine mist, or you can wipe the fitting and pipes once again, this time with a wet rag.

The joint is completed.

Now, take two caps. Polish both of them, flux the ends of the 6" and 4" nipples, and solder them in place. You will be sucking solder into the bottom of a vertical socket again, the cap on top of the 6" nipple. Half-inch caps are so small that it is difficult to direct your flame at the cap without hitting a great deal of the pipe itself, unless you direct it in such a way as to apply most of the flame to the very top of the cap. That's how I do it. The flame hits both the top and the sides, and, preferably, barely touches the pipe. If it does, no matter, the cap heats so quickly that you have the joint made while there still is a significant temperature differential between the cap and the tube end.

Wipe and quench the joints, as before. You are left with one unfitted end, the 12" end.

I've saved that end for the "test."

Take a ½" *male* adapter, polish the socket, remove your assembly from the vise or holder, and attach the adapter to the free end of the 12" nipple. Replace the assembly in the vise or holder, with the 12" nipple at an angle that is slightly above the horizontal. The end with the adapter should be slightly *higher* than the fitting. Reason: Whenever you solder male or female adapters, unions, or any other fittings having pipe threads, there is a possibility of the solder being sucked into the area of the threads by the capillary action. Once that has happened, it is almost impossible to clean them out completely, and, in most cases, they are useless. That can happen to a male adapter, if it is angled *below* the horizontal, by runoff. You put the solder into the cup or socket, and it fills up and runs out the end of the fitting, around the end and over the outside surface of the fitting, fouling the threads.

Wherever possible, solder threaded fittings in a holder or vise.

I'll tell you what to do, if that is impossible, a bit further on.

For now, let's get on with our test. Buy a ½" female pipe-thread x standard female hose-thread adapter, attach the pipe-thread end to your assembly, and the hose to the other end of the adapter, and turn on the water. If the assembly doesn't leak, you have done a first-rate job of soldering. If it leaks, here's how to remedy the problem:

First, make sure that all the water is out of the assembly. CAUTION: *If there is the slightest bit of water in a copper pipe, you cannot solder it. Reason: When you apply heat to a pipe containing water, you boil the water, setting the surface temperature of the pipe and fitting at the temperature of boiling water, 212° F at sea level. You require a great deal more heat to melt 50/50 solder.* If you're lucky, there is so little water in the assembly or pipe that you can boil it out, and get on with making the joint. If you're not, there are ways of overcoming the problem.

Right now, you must reheat the leaking joint. As soon as the solder begins to melt, take the flame away from the joint, and quickly apply flux around the place where the fitting joins the pipe. It will suck in, and brighten the surface that you have treated.

Apply a bit more heat, and some solder (not too much). You will notice the solder being drawn into the fitting. The joint is sound!

CAUTION: *The greatest cause of failed joints is overheating. When the joint sputters and sizzles as you try to draw in the solder, it's too hot. Sometimes, it will try to reject the solder rather than attract it. If the fitting turns a permanent blue color, you have overheated for too long, have probably ruined the fitting, and should replace it. If you have caramelized the flux around the joint (turned it brown and sticky-hard), you've used too much flux and have overheated the pipe, before getting the fitting hot enough to draw solder.*

HELPFUL HINT: One of the most useful tools you'll ever buy is a small five-and-dime-store mirror. With the mirror, you can examine each joint before you fill your lines with water. Just pass the mirror around the portions of fittings that you can't see easily. You should see solder all around. If you see gaps or "holidays," you know what you must do. If you see black spots that worry you, spray or clean the joint with water first and then look again. The spots could be flux. If you still see them, don't take a chance. Resolder!

When you are soldering threaded fittings in the line, I find it helpful to pursue the following procedure. The same technique is also effective with unions. Heat your fitting very carefully. Apply just enough solder to draw it into the socket or cup of the fitting. If you can look into the threaded end of the fitting, as soon as you see a silver ring appear where the tube joins the fitting *inside,* the joint is made. Again, you may see deceptive dark or black spots, especially at the bottom, which may cause you some anxiety. They are probably flux. Spray or wipe the inside with the tip of a water-soaked rag, and you will probably see them disappear. If they don't, just dig at the spot with the tip of a knife blade. If you still have a spot, which could be a small "holiday," simply reheat the joint in the normal fashion and watch the silver ring in the fitting until it closes. CAUTION: *Apply the heat gradually. Don't overheat! You cannot overcome soldering problems by applying too much heat.*

HELPFUL HINTS: One of your greatest problems will occur when you have to install dielectric unions. These unions have plastic insulating parts that can't take heat at the temperature required to make a good, sound solder joint. My initial suggestion is that you try to make the copper side of the joint onto a nipple, with all the plastic parts and the threaded end of the union removed. Later, when you make up the assembly, which may require the union to be assembled and in place, simply wrap the entire union assembly in a sopping-wet rag, and go right ahead. Remember that water boils at 212°F. The plastic will easily take that temperature, and it is impossible for the union to get hotter than 212°F as long as there is water in your rag. This same principle can be used elsewhere in your soldering work.

Looking for the silver ring when you solder dielectric unions is very important. You don't want to over-solder and have solder running over on the collar over which you must fit the union nut. The union nut has such a close tolerance that it is next to impossible to fit it up if there is solder on the socket barrel. If you goof, reheat the socket barrel, and as soon as the solder runs, wipe it quickly with a *dry* rag. This same technique applies to bronze ground-joint unions for which you may occasionally find some uses. With these types of unions, however, you don't have to worry about plastic parts. There are none.

I think the time and materials involved will be well invested if you tried to make up a series of sample assemblies to test your technique.

I have given you all of the fundamentals. Now all that's needed is practice. There is no substitute for that experience.

Try vertical and horizontal joints. Take joints apart and try to resolder them. You'll soon find out how difficult that is to do.

Speaking of resoldering joints, normally you apply flux only to the pipe or male side of the joint. If you want to attach a new fitting to a pipe which has residual solder on the joint end, flux the fitting after you've polished it. Heat the pipe to get the solder running, and while it is still melted, push the fitting onto it with your pump pliers (the fitting will get too hot to handle almost instantly). Then keep your flame in contact with the fitting and rotate it to seat it all the way. At this point, I usually give the joint another flux treatment to make sure, and finish in normal fashion.

SAFETY TIPS WHILE SOLDERING

Whenever you are working inside walls or near wood of any kind, try to place a metal or asbestos shield between the joint to be made and the wooden surfaces in the vicinity. Be sure that you have your fire extinguisher, water, or sand bucket handy. After you have soldered a particularly difficult joint, which may have caused you to touch wood and create embers on its surface, spray the wood with your water sprayer thoroughly, so that you can see the water running off and cannot see any glowing embers anywhere.

At the end of the day, go over every single joint you have made and check all adjacent surfaces to make sure that there is absolutely no trace of embers or other fire symptoms.

Whenever you are working in a crawl space or other restricted area, *do not ever* work with flame or solder right over your face. Always keep to the side so that, if hot solder or flux drops, it will not strike your body or clothing. Fire and burning injuries are the most prevalent in this work.

Don't take chances with your life, the lives of others, and the security of your home. Observe the strict safety procedures outlined in this book and go beyond them, if you can see other measures that would make the job safer.

Whatever it takes, you are obliged to do! Don't take shortcuts. Use common sense!

CONSTRUCTION TECHNIQUES

When installing copper pipe, you need to know that it should have a slight fall to the point where it enters the house. Not many people are aware of that fact, but those of you who live in freezing country in the northern tier of states should be especially conscious of this situation. You must have one or more low points to which to drain your water system if you are called away during the winter.

The drainage function can be easily accomplished through your hose bibbs and a gate or globe valve which you can install at the lowest practical spot in your water system.

You know the techniques for soldering. Now I'll walk you through the procedures for installation.

Right in front, it is important to know how high to take your water services to your fixtures, and to other bibbs, faucets, and valves.

The angle stops for kitchen sinks should enter the cabinet between 20″ and 24″ from *floor level.* They should be between 8″ and 12″ apart, with the sink drain in the center of the interval between them, if possible. I usually install kitchen-sink angle stops so that they will come into the cabinet around 11″ under bottom of the countertop, and between 9″ and 10″ apart. In that way, you can use the shortest possible chrome supply tubing between the angle stops and the sink-faucet inlet barrels. The code doesn't deal with this subject, so you are safe following common-sense practice here.

Incidentally, most kitchen sinks are set into cabinets whose tops are between 36″ and 37″ from the floor, depending on the thickness of the countertop. The most common cabinet inner-height dimension is 35″, which will receive nearly all of the dishwashers manufactured in this country.

The angle stops for lavatories (bathroom sinks) are normally set at between 20″ and 22″ from *floor level,* and are most usually 8″ apart.

The one cold-water angle stop supplying the water closet (toilet) is usually set at "six-square," or 6″ from the floor and 6″ from the centerline of the closet flange, on the left side as you look at the toilet. This is one of the only times in plumbing when the cold water line enters a fixture on the left side.

Showers and tubs are "hard-plumbed," which is to say that they do not usually have angle stops, but are piped directly to their working valves instead.

I do something that most plumbers don't. I always insert a pair of globe valves on the risers going to tubs and showers, if possible. Certainly, they couldn't be installed on lines under concrete as would be the case in a slab-on-grade house. But, by and large, if the piping is being installed in crawl spaces or in garage joists, the globe valves can be installed quite easily.

I think that it's important to be able to turn off any individual piece of equipment in your plumbing system. The only way you'll be able to do that is by installing the two globe valves per shower and tub.

Construction of the system is now just a matter of cutting and fitting lengths of pipe and fittings.

To install the house main from the meter, obviously you're going to have to excavate to the meter. This is one line which doesn't require the removal of the old pipe, unless it is convenient. Most of the time it isn't.

If one is available, rent a Ditch-Witch or other type of small trenching machine, and cut a trench, 8" deep, to the meter. If you are unable to obtain the trencher, you're back to your "Siberian teaspoon."

You should get in touch with your local water department and find out what type of fitting you should use between your main line and the meter. Tell them that you're using copper pipe, and they should have the answer. In fact, you probably won't be permitted to make the hookup yourself. They will most likely insist on doing that, or, at least, being there when the connection is made.

Now, make up your line to the entrance valve. Don't forget to install a threaded outlet tee in your line immediately before you install the main water valve. You're going to screw in your "special" hose bibb in that tee. This is the only garden faucet that will not be turned off when the main valve is closed.

Now, if you are in freezing country you should tee off just above the main shutoff and install your drain valve in another tee. This one is simply a tee with a short nipple and a globe or gate valve soldered in place, with a length of tube running to within 6" of the ground, which should have a pad of concrete to keep the soil from being washed away when you drain the house system. This valve should be located in such a place as to drain all the water out of the house lines.

If you'd prefer not to have so much hardware in one place, you can always solder your drain valve to a drop, elsewhere. Just make sure that, wherever it is placed, it can empty all the lines of water.

Also, in freezing country, any exposed copper must be insulated to the extent required by the local code and your knowledge of duration and levels of the freezing temperatures in your area.

Incidentally, some people prefer to have the main shutoff valve inside rather than outside, especially in areas troubled by vandalism. That's okay as long as the valve is accessible. Everyone in the house should be familiar with its location and operation. If it is behind a door, in a workshop or storage room, be sure that you have a plainly visible sign on the door saying *main water valve* or something like that. If emergency personnel have to shut off the water, they have to be able to find the valve quickly.

HELPFUL HINT: Whenever you drill holes for pipe, make sure that you have sufficient work room. Normally, you drill a ¾″ hole for ½″ copper pipe, and a 1″ hole for ¾″. One inch requires a 1¼″ bore.

Always try to lay your lines perpendicular to the studs or in them (the preferred procedure). Lines should be parallel or perpendicular to each other, to give a proper, finished, and professional appearance.

In the end, however, it's the way the lines do the job that's important. *Never forget that!* If you have a choice between appearance and function, choose function every time.

HELPFUL HINT: If you can, install your service drops or risers before you run the rest of your mains and service feeds. If you'd rather not, then at least mark each hot- and cold-water drop or riser with some sort of long marker or stick. Some people find it helpful to put little white-and-red rag flags on them. In this way, you can see where your hot and cold services are, with respect to the mains and feeds which you are running, as you install them.

With these ideas firmly in mind, run your pipe. CAUTION: *Be sure that you support the pipe adequately (one clamp or loop of plumber's tape every six feet, on the horizontal lay, and one clamp per ten feet on the vertical lay).*

Measurement between fittings: When you make measurements between fittings, measure from face to face, and *add* for the sockets, to give you the cutting length of your pipe. In their infinite wisdom, the inventors of this piping system have standardized the cup depths of all copper water-pipe fittings. Half-inch fittings have ½″ cup depths. Three-fourths-inch fittings have ¾″ cup depths. One-inch fittings have 1″ cup depths, and so on. The same rule does *not* apply to copper DWV (Drain, Waste, and Vent) fittings. To determine their cup depths for each size, don't take chances. Measure them with your tape. An interesting fact is that, however you measure distances, from the seat of a socket to the outside of the tee or elbow, you will always find it to be the nominal dimension of the fitting. For example: If you want to fit your perpendicular run of pipe tightly up against the beam you are approaching, measure the distance from the last fitting to the beam, and that's the cutting dimension of your pipe. If it's ¾″, you would add ¾″ that the elbow adds to the end of the pipe. *You don't gain or lose a thing.* If you are coupling lengths of ½″ pipe together and you want to hit a vertical water service right-on, put your ½″ elbow in place on the riser or drop, and measure from the horizontal face of the elbow to the face of the coupling, and add ½″ for each socket. Cut your pipe exactly 1″ longer than your face-to-face measurement.

HELPFUL HINT: Quite often, reducing fittings are not available in a full range of types. You can solve that problem by buying a dozen or so solder-type bushings, and an equal number of solder-type concentric reducers (just call them reducers). The difference between the two is that the *larger end* of the *bushing* goes into a *fitting* and the smaller end receives the smaller pipe. The *larger end* of the *reducer* is a socket that receives the larger pipe. The smaller end is also a socket just like the bushing. You can readily see how valuable those two fittings can be.

CAUTION: *The code does not permit either water services or drains to come up through the floor. They must be in the walls for a variety of reasons, not the least of which is simple protection of the pipe.* Also, *services in walls don't usually foster mop-water leaks to the room below. That can happen if the holes in the floor are not caulked securely around the pipe risers. The water simply runs down the outside of the pipe, and puddles in the ceiling below until it finally runs out of cracks that the leaks have been instrumental in producing.*

That means that your risers or drops will terminate in elbows, into which will be soldered 4″ or larger nipples. If you are going to sheetrock, *after you've finished your plumbing project,* the usual practice is to cap the 4″ nipple with a solder cap, so that you can test the lines later, before the services are hooked up to the fixtures. The caps make very neat, slender penetrations of the sheetrock. If you had attached your angle stops before sheetrocking, you'd have to make much larger holes in the drywall, not a very professional way to do things.

On the other hand, if you are in an undercabinet area, often you can terminate your service with an angle stop. CAUTION: *Remember to turn off the angle stop as soon as you have completed each service. If you don't, when you test your lines, you could end up in the middle of your own lake.*

HELPFUL HINT: *The only* type of angle stops to use is what we used to call "speedways." These valves have compression nuts on both the input and the output sides. Coming in is ½" copper-tubing and going out is ¾" or ½" OD tubing for your flexible chrome or plastic service connectors. The old-fashioned type had ½" female iron-pipe thread on the input, and compression on the output. With this goodie, you either had to use a *brass* threaded nipple (costs plenty!), or a combination of tubing and a male adapter (looks lousy!). Either combination restricted options.

With the "speedway," you are able to use type M copper right out to its small, neat cap, which makes drywall installation a breeze.

ANOTHER HELPFUL HINT: There is an even newer version of the old "speedway." This plumber's dream combines the angle stop with a permanently affixed accordian-type flexible supply. The supplies can be purchased in nearly all sizes from 9" to 24". I try to keep all of my supplies to 9" for toilets, and 12" for everything else.

A *last-resort gambit if you must resolder a pipe containing water*—not under pressure—*that is preventing the joint from getting hot enough to melt solder, is the old "bread trick."* Go to the store and buy a small loaf of plain, white, sliced American bread. The best way to tell if you have the right stuff is to take a slice and roll it into a tight ball, which it should do without crumbling. You should almost be able to bounce it on the floor like a rubber ball. Then you *know* that you have the right stuff.

After you've emptied the pipe of as much water as you can, roll a piece of this bread into a tight ball just a bit larger than the inside pipe diameter. With the eraser end of a lead pencil, or other similar device, push the ball into the pipe about 3", if possible. It must be in at least 2" to be beyond the immediate effect of the first intense heat as you make the pipe joint on a ¾" pipe. Also, the object of the exercise is to dam up the water as far away from the pipe end as possible. Of course, you would have prepared all surfaces and have your flux applied, before you inserted the plug of bread. You have about a minute from the time you placed the bread plug to make your joints, before you will get water seepage, if you're on the horizontal lay. If you're on the vertical lay, you may have less time. Once the job is done, don't worry about the bread. It will crumble and dissolve harmlessly in the water. I usually open a tap nearest the place where I used the plug, after I've made the repair, and most of it runs out at that point. Anyhow, it's only bread.

HOOKING UP THE WATER HEATER

Water heaters and water softeners are installed in much the same way.

At this point, I'd like to refer you back to Section Three where I explain water-heater installation thoroughly.

If you must replace your old one, be sure that your new water heater is at least a 40-gallon unit. I also recommend the heaters equipped with "energy-saver" pilot lights. In California you have no choice; it's the law.

It is important to note that your heater must be piped directly from your principal interior main. If you come into the house with 1" and split off to ¾" submains, then your heater should be piped in-and-out with ¾". If your house main and interior main are ¾", the same instruction applies.

There is no way you should run anything smaller than ¾" to and from any domestic water heater.

CAUTIONS:

1. *If the heater is installed in a garage or anywhere else where it might be subjected to gasoline or other flammable fumes and gases, you must mount it on a platform at least 18" high.*

2. *Every heater must have a temperature-and-pressure-relief valve (T & P). From the outlet of that valve you must extend a tube that is the full diameter of the outlet (usually ¾"), either to within 6" of a floor that drains to a properly plumbed floor drain, or you must pipe it to "daylight" (outside of the house).*

3. *The vent pipe must be the same size as the draft-diverter outlet. Up until recently, almost all domestic hot-water heaters had 3" vents. Now, more and more of them are coming out with 4" vents. If your new heater has a 4" vent, you cannot pinch it down to 3". You must install a new vent.*

4. *As mentioned in Section Three, you must use a UPC listed gas*

connector. Be sure that it has the rated diameter of the control inlet (usually ½"), and is plastic-coated.

5. You must use dielectric unions on both the cold-water-inlet and the hot-water-outlet nipples (spuds).

6. One gate valve of the same pipe size as the spuds must be installed on the cold-water side. I recommend one on the hot-water side as well.

7. In nearly all jurisdictions you are required to install an approved receptor or pan beneath the heater (they are often called "smitty pans"). That pan must be plumbed to a drain or to "daylight." The purpose of this receptor is to accept water from a heater that has developed tank leaks. Of course, you should replace such a heater as soon as possible. But, in the meantime, this receptor may save your floors, carpets, and the ceilings and furniture downstairs.

For the installation of all other plumbing fixtures, see Section Three.

Hose bibbs may be installed at any convenient height, but most of them are placed in the range of 24" to 40" from the ground. CAUTION: *Equip your hose bibbs or faucets with approved back-flow prevention devices.*

Good planning and layout; technique born of practice, observation, and evaluation; the wisdom to ask relevant questions; fortitude; and family support are all it takes to produce the best plumbing job you've ever seen.

Once again, congratulations!

PLASTIC SEWER INSTALLATION
PLASTIC PIPE IS EASY; SEWERS ARE HARD
THE RULES

1. All drain, waste, and the horizontal portions of vents must fall at the rate of 1" per four feet toward the point at which they enter the city sewer or your septic tank (¼" per foot). If there is an overriding consideration, the inspector may allow it to fall at half that rate and still satisfy the code, but you must have a "variance" from the inspector, or you must maintain the standard code grade.

Overriding considerations include the distance from your house to the disposal facility. If your house is six hundred feet from the sewer, and the sewer is only six feet from the surface, it's obvious you can't maintain the standard code grade.

If you have a rambling ranch-style house with your kitchen and two bathrooms in the rear, away from the required exit of the sewer on the street side, in front, especially if it's "slab-on-grade," you may have to get a variance.

The same can be said of houses with restricted crawl spaces or natural impediments that make the standard code grade prohibitive.

In any case, *before* you deviate from the standard code grade, you must have permission from your building inspector.

HELPFUL HINT: If you run into this sort of problem, I think it best that you consult a competent civil engineer for design advice.

Of course, this situation can only exist if you are forced to reroute your existing system for some reason, or you're adding on to your present building.

Obviously, if you're simply retracing your existing system, you have a precedent. If the grade was shallow on your existing system, since you are simply replacing what was already there ("grandfathering") there shouldn't be too much difficulty getting the inspector to go along.

If, however, you are adding facilities as you renovate, such as that spare bathroom you've been waiting so long to build, the inspector may take a different view, and require you to retain an engineer to get both you and him off the hook.

Whatever the situation, grade is an overriding consideration in sewer construction, for very obvious reasons.

2. The size of your main house sewer is determined by the number and type of fixtures you plan to install, but usually 3″ is adequate for most residences nowadays. If you have numbers of floor drains, and waste from outbuildings to consider, it is more practical to install 4″. In any case, if you have any questions on the point, ask your building inspector. He's the last word anyway.
3. The sizes of the drains for various fixtures are as follows:

A. Water closet (toilet)—3″ or 4″ (your choice)
B. Kitchen sink, with or without waste disposer and dishwasher—1½″
C. Shower—2″
D. Bathtub or bathtub-shower combination—1½″
E. Clotheswasher standpipe—2″
F. Bidet—1½″
G. Laundry tub—1½″
H. Floor drains—2″
I. Bar sink—1½″
J. Lavatory (bathroom sink)—1¼″.

4. Every fixture and individual drain must be trapped and served by a vent. The trap must go right after the fixture or drain, and the vent must be installed downstream of the trap by no greater than the following distances:
   A. Water closet (toilet)—The trap is built in to the toilet base, and the vent should not be installed farther away from the closet flange than 6 ft., measured along the actual pipe from the face of the closet-flange to the centerline of the vent.
   B. Kitchen sink, bidet, laundry tub, and bar sink—Install undersink traps. Vents may not be farther away than 3′6″ from the centerline of the trap.
   C. Lavatory (bathroom sink)—Install undersink trap. Vent may not be farther away than 2′6″ from the centerline of the trap.
   D. Clotheswasher—The standpipe is directly connected to its trap, and the vent which serves it may not be more than 5 ft. away from the centerline of the trap. The trap must be set at between 6″ and 18″ above the floor.
   E. Stall shower—The trap is connected directly to the shower drain, beneath the base. The vent may not be farther than 5 ft. away from the centerline of the trap.
   F. Floor drains—Since these drains are served by 2″ lines, the code would normally provide for venting within 5 ft. of the centerline of the trap that you are required to install on every floor drain; however, the rule is quite often modified by the building department, especially if you have architect's or engineer's plans on your job.

A few things are easy to comprehend. One of them is that you can't have a vent sticking up in the middle of a slab of concrete, 50-ft. square, and yet it must be drained into an approved drain and trap. The answer is obvious. The drain is vented as soon as possible, after the line has passed beyond the point where it could be fouled by the slab runoff.

HELPFUL HINT: Get engineering advice on drains to be placed in large slabs of concrete. Otherwise, simple common sense should see you through your task. Remember, the water from outside drains should be conveyed to your storm-water system. Inside drains connect to your building sewer.

5. Placement of vents: Vents are installed downstream from the trap. *Every* plumbing fixture in your house *must* be trapped and vented.

The reason for traps and vents in combination is *safety.* Sewer gas is *methane,* a highly explosive substance. So means were devised, early on, to control the entry of methane into living spaces. Vents must be located between each fixture and the city sewer or septic tank, so that the gas will have an easy way to get out of the pipes *before* it can exit through one of your fixtures. The trap is simply the device which plugs the pipe with water, immediately before it connects to a fixture. It blocks waste pipes, forcing the gas to back up and go through the vent. The two devices work together, and *must not* either be eliminated or disabled, or you are in real danger of methane-gas buildup, and possible explosion within your home.

It is possible to connect two or more vents together, to simplify their exit from your roof. In order to be able to determine size combinations, you should consult your local code; however, I can give you examples, with respect to two common situations:

The kitchen: If you have both a normal sink and a bar sink in the kitchen, they can both be vented into one 1½" through-the-roof vent.

The bathroom: One toilet, one lavatory, one shower, one bathtub, and one bidet can be vented into one common 2" vent.

As a matter of fact, if your laundry, bathroom, and kitchen are very close together, they can all be vented out in a single through-the-roof 2″ pipe. Additional bathrooms and other facilities will change that condition.

CAUTIONS: *No vent may be less than 1¼″ in size, or less than one-half the size of the drain to which it is connected.*

Be very careful of vent runs in excess of 50 ft. Check your local code carefully on that point. Before you can install your first horizontal *section of vent, you must be at least 6″ above the flood level of the fixture served by the vent. By "flood level" I mean the level at which the fixture will overflow its basin, bowl, pan, rim, or standpipe. If you must jog-over somewhere, to get up through the studs to the proper place of roof-exit or connection to other pipes, you must do it by ⅛-bend (45-degree elbow), if you are lower than 6″ above the flood level of the fixture.*

This can be a problem, especially on kitchen sinks with picture windows centered right above the countertop. In most of these cases, the horizontal vent lay is right under the window-frame, which could mean that it would be considerably lower than 6″ above the flood level of the sink. Before you install the pipe, check with your inspector. You will need a variance.

Where possible, try to change the direction of all vents with ⅛-bends (see p. 196)

If you have an island sink, you may be in for a bit of a problem. As always, however, there is a solution—the loop-vent (see illustration). This special type of vent is permitted by Section 614 of the *UPC*, but check your local code.

These types of vents are rather complicated, but effective for the use intended.

As you can see from the illustration, the vent itself forms a loop with the drain, serving the kitchen-sink trap. You must branch off the waste line with a combo (combination wye and ⅛-bend) (see p. 198), whose run (like the run of a tee), continues upstream to other fixtures or ends in a cleanout. I find that, with loop-vented sinks, a separate cleanout is a wise addition.

On the vertical 1½″ you cement a sanitary tee, with the flow channel down, to which you cement a 4″ or 6″ nipple, that you can temporarily plug with a rag.

Cement a nipple into the top of the sanitary tee. The nip-

Loop Vent Conforms to Section 614. Uniform Plumbing Code—1979, as amended.

Drawing of loop vent, rendered by the International Association of *Plumbing* and Mechanical Officials (IAPMO).

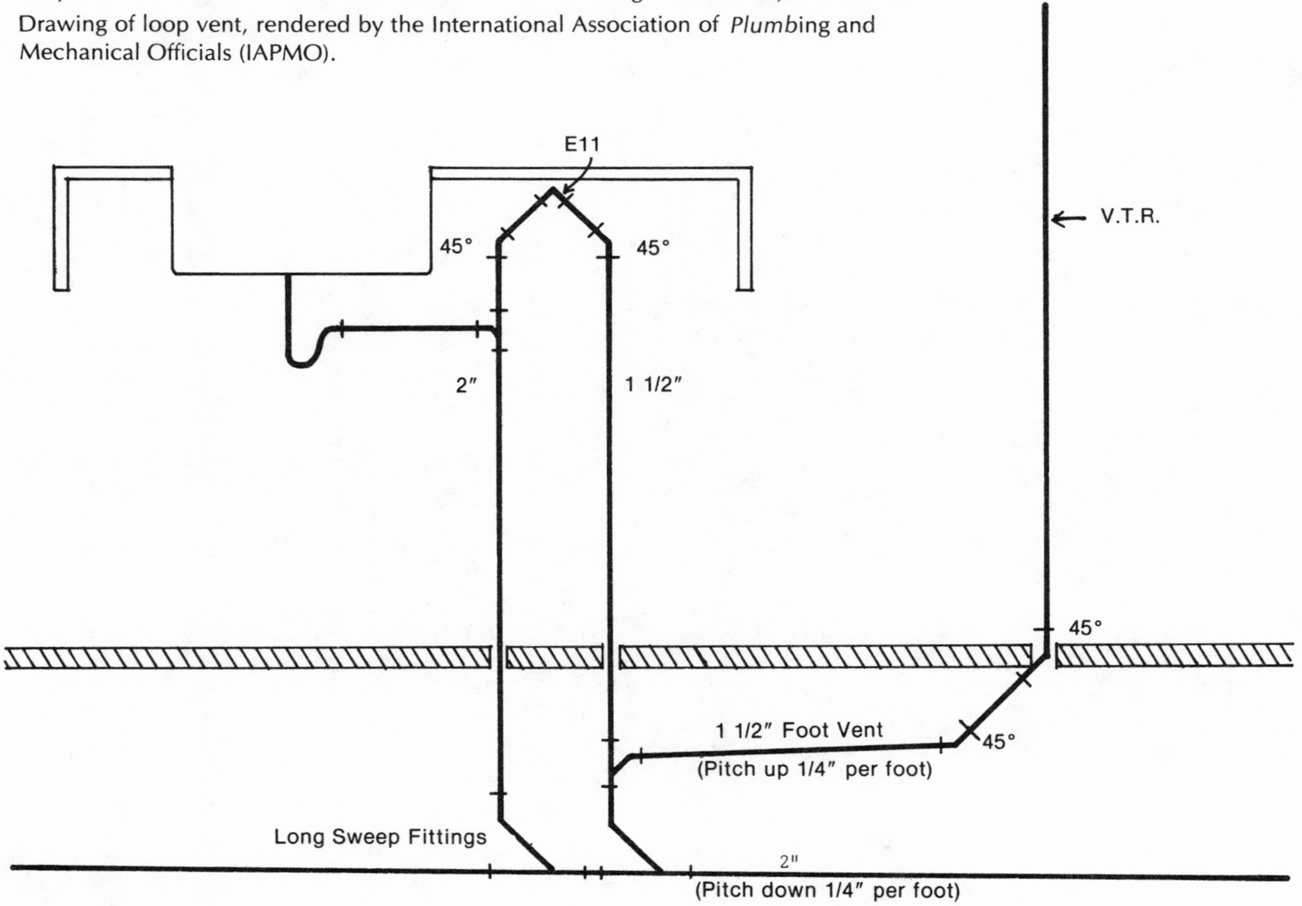

ple should be long enough to reach a point where the bottom of an elbow (¼-bend or 90-degree elbow) is no lower than the drainboard of your sink. That means that you'll have to build into the backsplash section of the sink, which should be large enough to accommodate a 1½″ pipe plus the additional width of the fittings. You then cement a short nipple to the ¼-bend. Cement a ⅛-bend (ideally) or ¼-bend (if absolutely necessary), to the nipple. The outlet should be looking down at the drain line. Cement a wye in the drain line itself, downstream from the sink drain, at a place where it can receive the pieces of pipe that you are going to install between the ⅛-bend or ¼-bend which I've just told you to cement to the top of the loop. If you come down from a ¼-bend, you're going to have to install another ⅛-bend in your descending line to compensate, so that you can interface with the wye that you've installed in the drain line. You may need additional ⅛-bends on the loop line to change direction, especially in the case of a waste line that is not running in the same direction as your sink counter. Immediately after you get under the floor, cement a wye and a street ⅛-bend (with the wye-arm up) into the loop vent, facing it in the direction of the nearest partition or wall to which you must run your vent line. Then take the vent under the floor to the wall and up, in the normal fashion. When you're through the roof, the vent is done.

All other vents are ordinary risers through the roof, either singly or in combination.

Now that I've thoroughly confused you with talk of ⅛-bends, wyes, and combos, I think that it's time to describe drainage fittings and how they work.

DRAINAGE FITTINGS AND FIT-UP

1/16-bend—22 ½-degree elbow
⅛-bend—45-degree elbow
¼-bend—90-degree elbow

Street elbows or bends have the female socket on one end and the male spigot (the same outside diameter as the pipe to which it connects) on the other end.

Sanitary tee—looks like a tee for water service but has a drainage-direction channel from the outlet into the run of the tee.

1. *1/16-bend (22 1/2 degrees)*
2. *Street 1/16-bend*
3. *1/8-bend (45 degrees)*
4. *Street 1/8-bend*
5. *Long-sweep 1/4-bend (90 degrees)*
6. *1/4-bend*
7. *Sanitary tee*
8. *Wye (lateral)*
9. *Combo*
10. *Long-sweep 1/4-bend with high-heel inlet*
11. *Reducer*
12. *Plug for cleanout*
13. *Bushing*
14. *Fitting cleanout adapter*
15. *Male fitting adapter*
16. *Male pipe adapter*
17. *Double sanitary tee*
18. *Pipe cleanout adapter (female adapter)*

Special Note: *Fittings are not to scale with respect to actual sizes. Small fittings are shown large to enhance detail.*

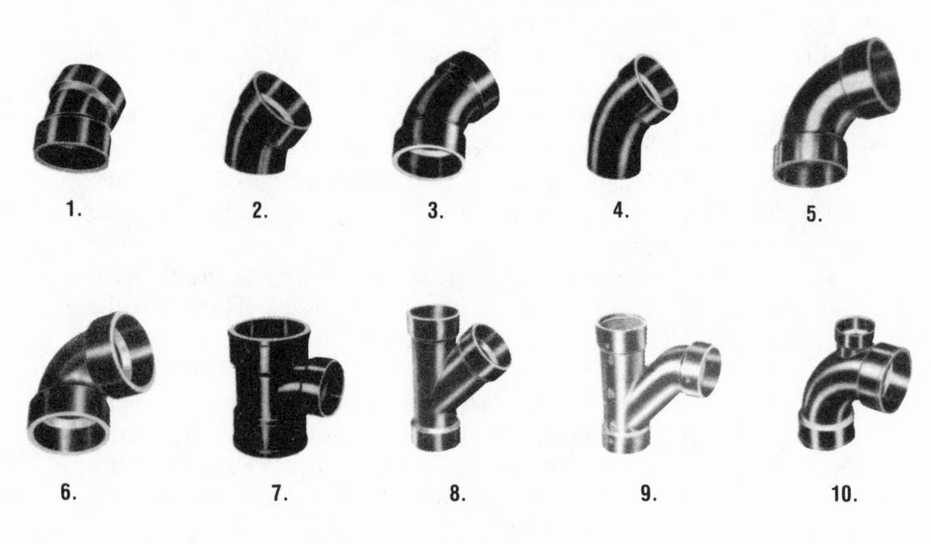

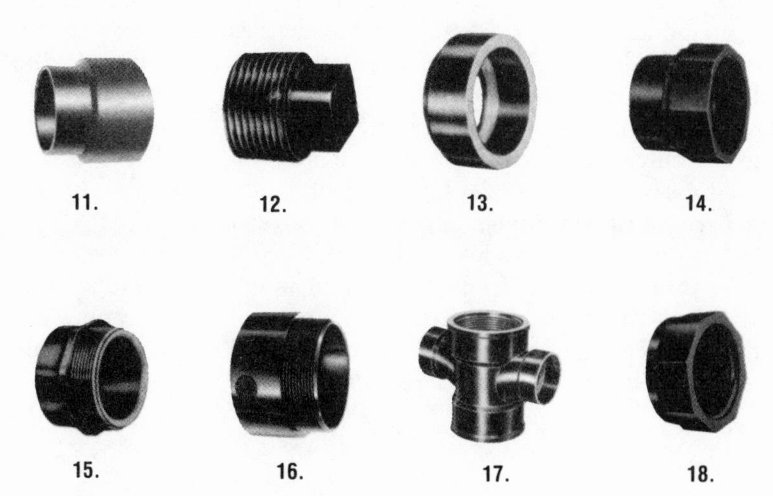

Drainage fittings

Wye—resembles a tee but the outlet is a lateral, 45 degrees from the perpendicular, which allows drainage to enter at an angle conducive to good flow into the lines, with minimum disturbance to drainage already flowing from upstream.

Combo (combination wye and ⅛-bend)—This fitting looks like a pipe with an elephant's trunk. There are many places where you cannot use the sanitary tee, by code. In those cases, the combo is the only way to go, when waste enters a main or submain at right angles.

Long-sweep ¼-bend (90-degree elbow)—This fitting is the equivalent of two ⅛-bends (45-degree elbows). This bend is used for actual waste or sewage, whereas the standard ¼-bend can only be used for venting. *Reason: The code prohibits the use of most fittings with an angle more acute than 45 degrees to conduct waste or sewage.* The old plumber's standby was a pair of ⅛-bends (one street and one standard) joined together. This long-sweep bend gives exactly the same sweep as two ⅛-bends together.

Reducers (bell reducers, concentric reducers) are shaped like bells and have sockets to receive pipe of different sizes. These are the only fittings which you are permitted to use to reduce the size of lines, other than reducing fittings (bends, combos, sanitary tees, etc.). You may *not* use bushings in waste or sewage lines. *Obviously you may not reduce any drain or sewer from larger to smaller in the direction of flow.* If you increase the line from smaller to larger, this is the fitting you must use. It should be noted that this fitting is often called a hub x hub (hub-to-hub) pipe increaser.

Bushings—These fittings accomplish the same thing as reducers, but may only be used in vent lines. Bushings are cemented into fittings to reduce an outlet or run. The outside is exactly the same size as the pipe diameter, and the inside is a socket to receive the reduced-size pipe.

HELPFUL HINT: More and more jurisdictions are accepting bushings in drainage and waste lines, but before you use them in that service, clear them with your local inspector. They are easier to use, and make for tighter installations in restricted spaces.

Cleanouts—The female-adapter-type is like a bushing, with female threads on one end. A threaded plug of the proper size screws into the threads to make an access point for cleaning out lines. The coupling-type cements to a pipe end and has the same threads into which the plug is screwed.

Coupling—As its name suggests, the coupling is designed to fasten two lines together, without reduction.

Male adapter—This one comes two ways, hub (socket) x male pipe thread and fitting x male pipe thread (to be cemented into a fitting).

SPECIAL COMMENT: The words "socket," "cup," and "hub" are interchangeable.

Female adapter—This fitting is similar to the male adapter but is designed to receive a threaded pipe end or male-threaded fitting.

Cap—You seldom use solvent-welded (cemented) plastic caps, except to terminate the stub of an abandoned line. CAUTION: *It is preferable to cut out the fitting of an abandoned branch line and replace it with a length of pipe and two repair couplings, either of the solvent-welded type or the sleeve-and-clamp type mentioned in Section Three (the segment on cast-iron pipe repair). Capped stubs tend to collect sewage.*

It should be noted here that there are now available mechanical combination cap-cleanouts that clamp onto pipe ends. These fittings are similar in appearance to the sleeve couplings and caps used on "no-hub" cast-iron soil pipe, which are permitted in some jurisdictions in lieu of conventional cleanouts and caps.

Trap adapter—This fitting also comes in two types; one cements to a fitting, and the other makes on to a pipe end. The adapter is the termination of your services to sinks. It is designed to receive chrome-brass or plastic p-traps, and consists of a hub-type or fitting-insert body, a plastic compression nut, and a polyethylene or rubber-type gasket.

Double sanitary tees, double combos (double fixture fittings) and double wyes are available, but the code is very explicit on their use and would require a case-by-case assessment of the particular installation before it could be determined whether or not these fittings would be legal. Mostly, their use is restricted to the installation of fixtures back-to-back and in situations where two lines enter the same stack. A rule of thumb on that type of installation is: whenever two lines come into a stack from opposite directions, they may enter a double sanitary tee (cross) only if they are two sizes smaller than the stack that they are entering.

HELPFUL HINT: Drainage and vent pipe sizes are as follows: 1¼", 1½", 2", 2½", 3" 3½", 4", 4½", 5", 6" and up in full inches. For purposes of calculating two sizes up from a 3" water closet line, refer to the list of sizes and see that 4" is two sizes up from your 3".

As a general practice, you should *not* enter any drain or waste line through a fitting having a greater angle than 45 degrees. The code allows 60 degrees, but from a practical point of view the only 60-degree fittings I use are wyes, which must always be installed vertically. That means that you should use ⅛-bends, standard 45-degree wyes, combos, and long-sweep elbows (which have the sweep of two 45-degree ⅛-bends) for drains or waste lines. Keep in mind that you want your waste water to come into the stream of already flowing water at such an angle as not to cause a backup of the water that is flowing from upstream in the line to which you're connected. It's as simple as that. The system is intended to be self-scrubbing, if it is properly designed and built.

Keep in mind that plumbing, like any other mechanical trade or craft, is governed by simple physics. Your supply water must have large enough pipe to supply all the fixtures in your house. Your drainage lines must permit the unimpeded flow of waste from every fixture in your house. Both functions should be capable of occurring simultaneously, without problems. That's what the code is designed to allow, and that's what should be in the top of your mind, the moment you start to think about doing the job.

The closet flange is deserving of full discussion. If you set it correctly, you've got it made. If you screw up the installation, you could tear out your fit-up and have to do it over from where the closet elbow attaches to the drain, which can be a total pain in the rear.

Closet flanges come in two types, all-plastic and plastic with a metal-flange, which can swivel on the plastic core that attaches to the closet elbow. The metal-flange type is the most convenient, because you can adjust the bolt slots to precisely the correct angle before you fasten it to the floor.

CAUTION: *Because this type is made of relatively thin steel, some inspectors frown on it. The possibility of rapid corrosion exists, but since the actual seal is to the plastic, those fears appear to be based upon excessive caution. Anyway, it's best to check it out before you install these types of closet flanges.*

Closet flanges also come into two sizes, 4″ and 3″. At this time, I think it best to mention *closet bends or elbows,* since they are the next fitting down the pipe from your flange.

SPECIAL COMMENTS: Closet bends usually have a shorter sweep than would occur if you used two ⅛-bends, but this is allowed by the code as being expedient in normal construction. If they didn't have comparatively short radii, you'd never get them to fit properly between floors and ceilings. They'd stick down too far. If you have the space, as you would have if you were plumbing in the joists of a garage, utility room, or similar space where appearance isn't all that important, I find it better practice to drop down into a combo, and install a cleanout right there, on a line to the rear of the draining line. *Of course that will only work if you have enough fall to the line into which you will deliver the waste.* You'd be surprised how easy it is to rod-out (unblock with larger equipment) a clogged line when you have a straight shot made possible by such an arrangement.

Closet bends or elbows come in two standard sizes, 4" and 4" x 3". If you are running 4" sewers to your water closets, the 4" is the proper flange for the job. The 4" x 3" is designed for 3" closet lines.

CAUTION: *When you are buying your closet flanges, I suggest that you buy straight 4" and use the closet bend or elbow to reduce to 3". It makes a much better flow pattern in the lines, and tends to resist clogging better than a 3" into a long-sweep elbow, which is the only ¼-bend recommended for 3" flanges.*

You can purchase closet flanges and closet bends in hub-to-spigot, spigot-to-hub, and hub-to-hub arrangements, which permit you to install the flange to the bend directly, or with a short length of pipe between the hub and the hub of the bend, to your horizontal drain.

CLEANOUTS—WHY AND WHERE

The use and placement of cleanouts is often a matter of code interpretation. Some inspectors are "cleanout happy." Others call for "bare bones" observance of the code.

Whether or not you should install a cleanout is a matter of the code, plus common sense. The last factor is the more important, by far. After all, you're the one who's going to have to maintain your plumbing system, and the judicious placement of cleanouts can make the job much easier.

The code says that every horizontal drain line over five feet in length must have a cleanout at its upper end. If your pipe slopes 28 degrees from the horizontal or more, a cleanout is not required. But that is a heckuva slope; even there, it isn't a bad idea to have a cleanout at the end of the line.

The rules get a bit more technical when dealing with the "house drain" and the "house sewer." The "house drain" is the main and submains in your house. The "house sewer" is the part that goes from your house foundations to the city sewers or your septic tank.

The code excepts drains in floors above the first floor from the cleanout requirement. Here again, it is a matter of inspector interpretation. Would the first floor be your basement or cellar? If the answer is yes, then crawl space beneath the house is the equivalent of a first floor. If the first floor of the residence itself, containing the primary living space, is considered the first floor, that would mean that you would only have to have cleanouts in your basement, and the floor above would be exempt.

There is a simple resolution of this problem. Put cleanouts in both places. Drainage piping in the finished ceilings and floors of your living space are exempt, since access for cleaning out drains is provided at each fixture location (see Section Three).

The ceilings of most basements and cellars are open joists. That gives great access to any piping, and is a logical place for cleanouts.

If it is impossible for you to install a regular cleanout at the high end of your "house drain," then you are required to install an approved two-way cleanout just inside your house, at the place where it joins the "house sewer." As you can see in the drawing, it has "rodding channels" going both ways, which permits you to put your snake in the top and make it go in either direction to clean out the line.

Common sense tells you to try to place your cleanouts where you can have easy access to them. It is obvious that you shouldn't install a cleanout up against a beam or joist. You'd never get a snake into it. In fact, you'd be lucky to get the plug or cap off. The code says that 2" and smaller cleanouts must have 12" of clearance. Larger sizes must have 18" of clearance. If you have to install a cleanout in the floor, it has to come up to floor level, and must be within 20 ft. of access, whether that access be a crawl hole, trap door, or other similar means of entry. The code also allows you to take underfloor cleanouts to "daylight."

Cleanouts in the middles of drains must enter the pipe stream at a 45-degree angle, so that you can snake the line from the upstream side of a blockage.

Wherever possible, put your cleanouts in place at the high ends of lines, and don't be skimpy about them. It is far better to have too many than too few.

IMPORTANT CAUTION: *Whenever you enter any drain from any other drain you must do so at the required grade, and the inlet of the receiving drain must be above the horizontal, at the code grade of 2 percent or around 88 degrees from the vertical. This applies to cleanouts as well.*

If you install any cleanouts in slabs, make sure that you place them under access plates of strength and durability.

ANOTHER CAUTION: *Whenever the fittings in any horizontal line total over 135 degrees in horizontal change of direction, the line must have one additional cleanout. The code doesn't say where that cleanout should be, but common sense tells you to examine the nature of the direction changes and install the cleanout in a place that will make it easiest to keep your drains clear, if rodding becomes necessary.*

The best fitting I've yet found for middle-of-the-line installation, to grade, is the combo, with a standard solvent-welded (cemented) cleanout at floor or deck level.

PLANNING

Actually this phase occurs so late in the plastic drainpipe segment of the book because I wanted you to get a feel of the nature of the undertaking. Onward now to the topic of *successful completion of the project.*

The construction portions have been written with renovation in mind. Not that you can't use the information gleaned from *Plumbing For Dummies* to give you sufficient background to plumb new construction, but, frankly, I can see a lot of frustrated weekend mechanics vowing to do-it-themselves, and feeling that they have enough information from this book to make it from start to finish.

I honestly hope that is the case. Nothing would make me happier. And I have tried my level best to point my remarks and details in that direction. But it will take an unusually adept, mechanically talented person to plumb a brand new house from scratch, even with this book, if she or he hasn't been around plumbers at work before. It's one thing to read about the work, decide you can do it, then take the tools in your hands and actually get it underway with the prognosis of success, and quite another to succeed.

I suggest renovation first. Replacing old, slow-running, sagging drains with plastic. Perhaps you are anxious to try your wings, putting in that extra bathroom you've been planning for so long. I believe you can do that with the information you get from this book.

When you plan your pipe lines, draw them all out on a plan of your house, on which you've already indicated your existing plumbing. Put sizes to everything, so that your perspective will be complete.

If you have some basic drawing skills, show the new addition or proposed area of construction in section (side and end views).

Now, indicate with arrows the flow directions of the existing pipes. With that information, you should be able to figure out where you're going to cut into the existing cast-iron or plastic main.

You must decide whether you are going to use ABS or PVC. ABS is available in almost any store, from crackerbarrel general stores to the largest building-materials outlets in the nation. It is the least expensive approved drainage pipe.

It is said that ABS's biggest drawbacks are that it is not self-extinguishing in a fire and that it tends to support combustion. Well, that may be so, but then, I've seen houses that were plumbed in iron, steel, and copper, that weren't worth a thing after the fire had burnt out. I've seen plastic- and copper-plumbed houses that have come through some pretty rough fires relatively intact. The plastic was ABS.

PVC is prettier (it's white), it's stronger, and, for some reason, seems easier to handle. It is also considerably higher priced, and has the virtue of being self-extinguishing.

CAUTION: *Because PVC is a bit less in demand, most suppliers don't carry PVC if they stock ABS. If they do have PVC, they usually have it for water service, and the fittings are water-pattern pressure fittings which cannot be used for drainage. As a matter of fact, the only PVC water fittings that can be used in your waste system are the elbow and the tee, and then only in the vent system, 6'' above the flood line of the fixtures to be served. That is a code violation, but many building departments allow this one variance, almost automatically.*

I'm not putting PVC down. Actually, it is the better material, but, because of the economics involved, it is usually difficult to come by, especially in the fittings department. If you have the money to spare, and can locate a good source of PVC pipe and DWV fittings, go with it.

I honestly like to work with it in preference to ABS. The solvent cement is clear rather than black, and it generally makes less of a mess than ABS. But when it comes right down to the nitty-gritty, most of you will end up buying and working with ABS. It's a fine material, and, compared to laying cast-iron, it is heaven-sent.

As you are making your materials list, don't forget to include sufficient cleanouts. Realize that if you are going to cut into cast-iron soil pipe for your new service, you will have to buy a cast-iron wye (lateral), combo, a bend, or long-sweep elbow to make your entry. CAUTION: *You cannot cut out a section of cast-iron soil pipe and insert a plastic fitting to make the transition easier.*

Of course, you'll use a no-hub cast-iron fitting, so be sure you get enough sleeve couplers to make all connections.

If you are cutting into a plastic pipe line, there are also sleeve couplers available (the screw-tight stainless-band-over-the-neoprene-sleeve type, similar to the cast-iron no-hub coupler).

For the transition from iron to plastic, there is a special coupling called the "Calder coupling." Other manufacturers make couplings for any and every drainage transition: clay-to-iron, iron-to-asbestos-cement, plastic-to-iron, and all combinations of these and others. With these facts in hand, make your drawings, material list, and permit application (if you decide to seek the professional help of your local inspector).

DRAIN, WASTE AND VENT CONSTRUCTION

First and foremost, you must learn how to cut, fit and solvent-cement plastic pipe. It must be cut off cleanly and square, deburred both inside and out, and then coated evenly on both the outside of the pipe and the inside of the fitting. You must give the pipe a quarter-twist as you join the two cement-moistened surfaces.

I usually cut my pipe in a jig or miter box, with a Sandvik saw (see tool list), and deburr it with my thumbnail which I scrape around the end, both inside and out. Occasionally, I get a plastic splinter or two under my nail, but that is a small price to pay for efficiency and speed.

There are some rather expensive pipe cutters for plastic, and they work well. The best I've seen are manufactured by the Ridge Tool Company (Ridgid).

When you get ready to join the pipe, make sure that you haven't had a gab session with someone and allowed the joints to dry out. CAUTION: *Always check before you put the pieces together. If you have allowed the glue to become partially dry, you just won't get the pipe to go home into the fitting seat. You'll probably end up cutting it out and patching with a short pipe section and two couplings, or some equally costly and unprofessional-appearing method. (As you've probably surmised from my familiarity with this method, I'm occasionally plagued with making unprofessional remedies of my own stupid goofs.)*

I usually work from the fixture to the main. The first thing I do is figure out how much fall I'll have to have to meet code. That will usually determine where I'll enter the house main or submain (also referred to as the "house drain").

If my new bathroom is sixteen feet away from the nearest practical access to a main, I know that I'm going to have a fall 4" from the closet bend to the main. If the main terminates high, obviously I'm going to have to take my new line to a point farther down from its termination, to the place where it has fallen enough to furnish my new service sufficient fall to meet code.

A huge problem looms on the horizon, if the nearest access is in between joists, and I have to run under the joists to make any connection to it. Chances are that it's out of the question to try to get in at the nearest point. I then look for a vertical waste line from above, in the case of a two-story house. If there is none, then I will probably end up running my new line to a place near where the "house drain" joins the "house sewer" on its way to the city main.

The only other solution is to run your line down, and then outside of the house, underground, to a place on either main with enough fall to meet the code.

Sound complicated? It really isn't. It's just part of laying out the job *before* you start doing anything definite.

Let's do a typical bathroom, which embodies all the principles of kitchen drain and vent installation:

Our bathroom will have a water closet, lavatory, a bathtub, and separate shower.

The shower is in one corner. The tub is on the opposite wall. The toilet is next to the shower, and the washbasin is between the shower and the tub on the wall.

Obviously, it is most practical to vent the whole room in common.

For purposes of this illustration, we will assume that we are working to studs and joists, that there is no finished wall, and that there is a generous crawl space beneath the room.

We first set our drains, starting with the largest, the toilet. We drop the closet flange in place, and it rests in the hole which has been cut in the floor. There should be a hole in the floor for the shower and tub, as well. The hole for the closet flange should be 5½″ in diameter. Cut out a 4″ circle in the center of the showerpan area. The cut for the tub should be rectangular. At the drain end, cut a slot 6″ wide that will reach from about 2″ toward the back of the tub from the centerline of the drain in the bottom of the tub, to the wall at the drain end. If the joists interfere, you may have to notch them on top to permit the drain to nest in its proper place when the tub is lowered into position.

CAUTION: *With respect to the installation of toilets, they must have 30″ of clearance, 15″ from centerline to either side, so that fat people or folks with great big shoulders will have enough room. They must also have 24″ clearance from the front of the bowl to any partition, fixture, or other obstruction.*

Now, "dry-fit" the closet bend into place beneath the closet flange. If your drain will run parallel to the joists, and you'll have enough fall to pick up the other fixtures, by all means close-couple the bend to the flange. If you'll be running across the joists, then fit a short piece of pipe (nipple) to the flange and then to the closet elbow. Remember, if you have a straight 4″ flange and a 4″ x 3″ closet bend, the pipe between the flange and the bend will be 4″. If there is enough drop for the nipple, then there would be enough drop for a 4″ x 3″ closet flange to make onto a 3″ long-sweep 90-degree elbow. That will eliminate the need for any 4″ pipe.

If it would be possible to go to "daylight" (run a pipe from beneath the new bathroom out through a nearby wall), then you may consider draining your water closet into a combo, and taking off from the upstream side of the run, with a length of 3″ pipe. As soon as you get outside, fit a cleanout to it, and you will have a wonderful, easy point from which to unblock the bathroom line, if the need should arise. Otherwise, I usually take off from a wye in a new submain like this, very close to the toilet discharge point.

The next fitting on the downstream side of the closet bend or long-sweep elbow is a 3″ x 2″ or 4″ x 2″ sanitary tee, wye, or combo (depending on what size bathroom submain you have decided to install). This new fitting will take the water-closet vent. The flow channel of the fitting should be oriented exactly as if waste were being discharged into the wye. The idea is to exhaust the gases which flow toward the house from the city main or septic tank.

The closet, tub, and shower vents all break the code rule against not installing a horizontal vent line below 6″ above the flood level of a fixture. It is quite clear why. Venting below the floor level is the *only* way you can vent them at all.

Be sure that you have a definite grade to the vent as it runs to the point where you want to take it up through the wall to the roof. A 2″ line requires a 2⅝″ hole. If you want to get a coupling or the hub of an elbow pushed into the hole, you must drill a 2⅞″ hole. To carry the vent to a convenient place to join the other vents, go into the attic or ceiling crawl space of your new room. If it doesn't have either, then you're going to have to run your pipes through the studs, which means a lot of careful drilling. Using the rule of 25 percent of the board width in bearing members, your vent wall should either be built of 2″ x 6″ plates, studs, and sills; better yet, double up, with two 2″ x 4″ stud walls, set 3″ apart so that you can rise easily with your 2″ closet vent, and make the rest of the vent runs quickly and efficiently.

HELPFUL HINT: With your toilet next to the shower, and the lavatory on the same wall as the shower, be sure that the tub outlet is facing the same wall, so that you aren't in a situation of having to build more than one "special" wall.

While you're at it, think about your underfloor drain network, with respect to the bathroom submain. It is always best to run one straight line to the next major junction with either another major submain or the "house drain." That means that each of your fixtures has to drain into the submain from branches that are organized to enter at 45 degrees, and that, if the submain runs down a wall line, your cleanout will simply be an extension of that submain, rather than come out the back of a combo into which the toilet drains.

CAUTION: *Never run waste water through a closet connection. It would be possible to do so with any number of fittings available from the manufacturers, including the combo, long-sweep ¼-bend with either low or high heel inlets, which are code-approved in the 3" x 3" x 2" size, and that are used by some plumbers to simplify venting.*

Some jurisdictions will accept heel-inlet combos for venting purposes, but no one allows them to be used to receive waste water, which would then flow right through the closet discharge, a big no-no.

The suggested combo is intended to allow an easy cleanout connection. CAUTION: *Cleanouts and their lines must be the full size of the lines they serve.*

Of course you are permitted to enter any closet line downstream of its connection to the submain.

At this point, you should have "dry-fit" (joined without cementing) the elbow to the closet flange, your vent fitting, the length of pipe to the place where you will rise through the "special wall," and you should be ready for the shower connection.

HELPFUL HINT: You have two options now. You can enter the line that runs from the water closet to another submain, or enter the submain itself. Choose whichever is closest. Both the closet line and the submain will be no less than 3". If you want to run your shower line at right angles to the closet line, all you need to do is use a combo where it comes into the 3".

As you remember, the shower drain is 2". If you are going to have a tiled shower, all you must do is place a flanged tile-in drain fixture on top of the next assembly below, which will consist of a 2" plastic p-trap and a nipple into the drain fixture. CAUTION: *When you "lock up" the job (cement all your fittings in their final positions), you always leave this type of shower drain loose and uncemented. That gives the tile setter the required height options.* If you are setting the tile yourself, you will be grateful for your own foresight.

ANOTHER CAUTION: *You do the same thing with sink drains. Always leave a rag-stoppered nipple sticking out of the wall 6" or more. You can always cut the nipple when you fit up your sink, but when the sheetrock is up, you're stuck if you've made the connecting pipe too short.*

In the case of one of those plastic "unit" shower stalls, or with a molded-plastic shower-receptor pan, you can simply make up one of those plastic cement-type shower-drain fittings to the floor-drilling of the shower pan. CAUTION: *Be sure that you put a roll of plumber's putty around the flange of the drain fitting before you cinch it down, or sure-as-shootin' the thing will leak like a sieve.*

Every amateur tends to take the easy way out, if he or she can perceive a good and valid reason for doing so. You look at your rapidly developing project, beneath the floor. You realize that all of the underfloor vent connections from the shower, tub and toilet, *could possibly join each other under the floor, to simplify your venting through the wall. What a temptation.*

CAUTION: *Don't you do it; don't even think of it! It's the worst possible hazard with respect to drains backing up. I can't emphasize too strongly that you must wait until you are 6" above the flood level of the highest fixture in your bathroom before you can make such a connection. That fixture, incidentally, is the lavatory (bathroom sink).*

As an interesting aside, a friend of mine, Captain Murdoch Gillespie, Royal Engineers (Reserve), recently consulted me in connection with a plumbing renovation for his house in Petaluma, California. He referred to the vents as "stench pipes." I couldn't believe my ears, but, I reckon, that's what proper Englishmen call them.

I usually join vents near the ceiling, if there's no attic or ceiling crawl space, which is my preferred location.

HELPFUL HINT: If you're planning your addition at this time, by all means include such an under-roof area. It will eliminate the need for the double wall, since all of the vents will rise straight up into the attic, and needn't be run through studs, in the absence of the double wall. Also, you have more options. You can place your fixtures without regard to trying to locate them in such a way as to vent up one wall.

You have your plastic p-trap in place beneath the shower drain. The next fitting (on the horizontal lay) is your 2" x 1½" wye, combo, or sanitary tee (the latter being the least expensive and tightest fit), from which to run your 1½" vent line. Again it goes to the "special wall" and up.

HELPFUL HINT: The elbow from which you rise through the wall can be a standard 90-degree elbow. It needn't be a long-sweep type.

To plumb-in the tub, you should first assemble the waste-and-overflow fittings, which you must purchase separately from the tub itself. As soon as you've assembled them *according to the instructions*, you will see why I had you cut a slot in the floor for the tub fit-up. The horizontal drain tube that goes from the tub drain to the front of the tub, to join the descending overflow tube in a cast-brass tee and tailpiece, must nest somewhere. In some tubs, the tub frame or skirt is high enough to keep the tubing from hitting the floor. In others it isn't. And, besides, you need access to that entire assembly from beneath, to service it. Of course, if the tub is installed above a finished ceiling, the slot is only a convenience during the initial installation. You attach the tub-drain tailpiece to the plastic p-trap by means of a 1½" fitting trap adapter, mentioned earlier.

CAUTION: *Here again, when you install the tub drain, be sure that you putty it in place, in the usual fashion.*

SPECIAL COMMENT: It is worthy of note that many of the plastic p-traps that attach to both the tub and shower drains, are made with unions that are designed to swivel where the "J-section" joins the elbow that carries the waste into the horizontal lay.

HELPFUL HINTS: You can build your own p-trap from one standard-radius 90-degree elbow and two standard-radius 90-degree street elbows.

The L.A. pattern p-trap with union is available for the shower and tub drains. The p-trap has a slip joint on one end (which eliminates the trap adapter) and the usual hub or socket on the other, to take the horizontal drain. This p-trap is useful if you don't have to extend a tailpiece from the drains to the p-trap to permit you to take your horizontal drain under the studs below.

CAUTION: *The code prohibits any tailpieces in excess of 24", measured from the face of the drain fitting to the top of the p-trap tubing that forms the loop of the "J."*

To refresh your memory, a tailpiece is the portion of a fixture that extends down from it to the p-trap. The toilet has no tailpiece since its trap is cast into its base. The tailpieces on sinks of all types are usually chrome-brass extensions from the sink bottoms to chrome or plastic traps. The tailpieces on tubs and showers are either metal or plastic nipples joining their drains with p-traps.

The reason for this regulation is that too long a tailpiece may allow so much velocity in the falling water passing through the p-trap, that it will suck the trap dry, and cancel out a vital safety feature.

CAUTION: *A p-trap is useless unless it contains a proper water seal. That is why the code requires every drain, no matter where it is (including floor drains), to be served by a water source.*

When you are away from your home for awhile, it is a good idea to have a neighbor or your housesitter turn on a faucet over every drain in your house, at least once every few days, to keep the trap seals intact. Water evaporates, and when it does in a trap, it becomes a possible means for methane gas to enter your house. A lit match, dropped in a drain, can result in an explosion of incredible force!

Once the tub's trap is in place, you run your 1½" drain to the submain or closet drain, whichever is most direct, and you run your vent to the place where you intend to make your vertical lay up the wall, all of which must follow the instructions I've given for the construction of the shower drain.

The last item is your lavatory. It's the simplest, in that the trapping is done in the finished room. Drop a straight 1½" pipe from a sanitary tee with its flow channel down, to which you have fixed a 6" or longer nipple in its outlet, that will later receive the sink trap. If the base of your drop points directly at the submain or closet drain, simply install a 3" x 3" x 1½" combo and you're in. The lavy-line goes right into the 1½" long-sweep 90-degree ¼-bend or one or two 45-degree ⅛-bends, point them at the nearest available main and take them into the wye or combo on the horizontal lay.

Your vent is a snap. Simply install a pipe in the top of that sanitary tee. That's all there is to the vent.

SPECIAL COMMENT: Where I have assumed that you are plumbing your water closet, submain and main drains in 3", if you are using 4", simply substitute that size for the 3" called out.

CAUTION: *In setting your vents please don't forget the maximum distances from traps to vents, which I have mentioned earlier.*

COMBINING AND TERMINATING VENTS

At any point in the vertical lay or rise of a vent, you can jog it over by using ⅛-bends. I repeat, you are only permitted to use ¼-bends to move a vent pipe horizontally, when you are 6" or more over the flood level or rim of the fixture served by that vent.

On all vents you may use standard-pattern ¼-bends (short radius).

You may combine all of the 1½" vents in one bathroom, provided it is not back-to-back with another bathroom, and you try to combine all of those 1½″ vents with the original bathroom. Double bathrooms containing two of everything fall into the same category. Two full bathrooms consisting of two tubs, two lavatories, two water closets, two bidets, and two showers can be served by one 2″ vent.

With this in mind, suffice it to say that you are permitted to connect all of the 1½″ vents in *each* bathroom together. Then you can combine them with a 2″ manifold and take the vent to "daylight."

The water closets *must each* be served by 2″ vents, which can be joined to the 1½″ vents.

If your final through-the-roof 2″ vent will be located approximately in the middle of the interval between fixtures at both ends of the room, with one fixture immediately below that point, then you take ¼-bends and sanitary tees to the place you plan to rise, and install a 2″ double sanitary tee (which resembles a cross) with the branch flow patterns upward. The bottom socket or hub takes the middle fixture's vent, the two horizontal hubs receive vents from both extremes, and the top one holds the vent pipe which you will take to "daylight."

If your vent will be located at one or the other extreme, or anywhere in between, you'll have to use either a double sanitary tee (if a fixure is located immediately below the point of exit) or a sanitary tee anywhere along the horizontal connecting lay of the vent line. In all cases the flow patterns of fittings must be toward the top of the vent on the roof. The gas flow must be up and out, into the open air. With your drainage system, the water must be down and into the city sewer or septic tank.

CAUTION: *Be sure that all vent lines rise to the point of connection of the fitting to the final leg of the vent through the roof.*

The vent pipe must go through the roof and extend 6″ above the surrounding roof surface. Also, it can't be closer than 1 ft. from any vertical surface affixed to your roof, such as a false front or the curb around a sunken roof.

Where it comes through the roof, it must be "flashed" with an approved roof jack, which must be sealed to the roof with an approved mastic or bituminous sealing material.

CAUTION: *If you have a roof that is unusually configured, or that has a number of chimneys, heating vents, and other plumbing vents already in place, consult your local building department on the code requirements of any new vent. Also, you are not allowed to terminate your vent within 10 ft. of any window, for obvious reasons.*

Warning: It is strictly against the code to install your vent in plastic outside the building, exposed to sunlight. If you decide to replace an existing iron or steel vent, you may install the portion that's inside the house in plastic, but outside it must be metal. I suggest copper DWV. *(The 6" of vent extending up from your roof is exempt from that restriction.)*

Both ABS *and* PVC *are subject to ultraviolet attack (deterioration). If you still persist in running a full vent outside of your house in either of these plastics, you must paint them with latex paint, to protect them from this problem.*

Turn to Section Three for information on the installation of fixtures and appliances.

Your last caution in this section has to do with the matter of supporting all drain and vent pipework. There are galvanized-steel pipe clamps, standard pipe hangers, and galvanized-steel plumber's tape, all of which are satisfactory for such work. Some manufacturers are now marketing pipe hangers made of polyethylene and other similar materials. Some are acceptable under the code and some aren't, so I suggest you ask your inspector or consult the local code. In any case, your plastic sewers, inside the house, must be supported at least once every four feet on the horizontal lay. On the vertical lay, I recommend the same clamping schedule, even though it is not spelled out in the code, except to say that vertical lines shall be supported to maintain a straight alignment.

If you've followed instructions carefully, your bathroom and, later, your kitchen should be ready for inspection. You've won! I'm proud of you.

Typical Bathroom "Rough" Plumbing in Place

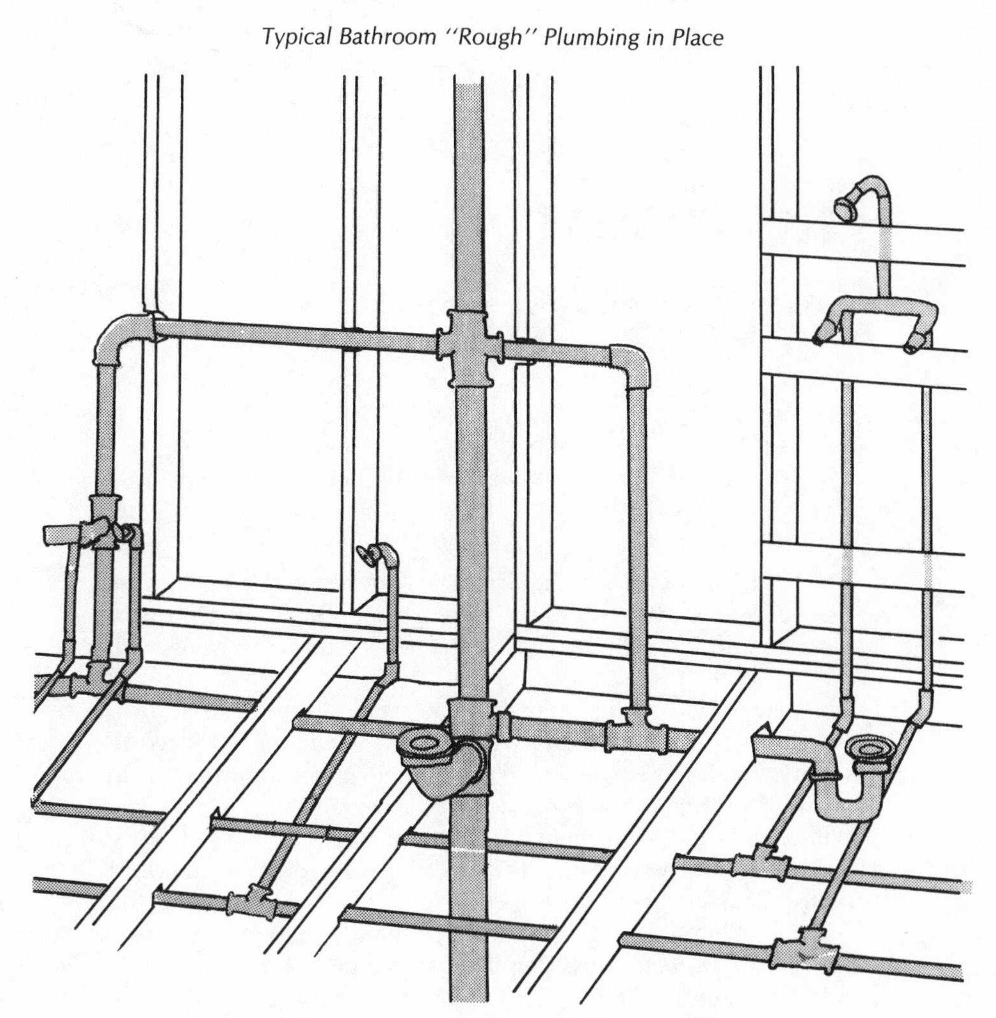

Completed rough house-plumbing system, showing both water-supply and drainage fittings in place.

SECTION 6

The Solar Connection

If every home in the United States of America and Canada were equipped with a basic hot-water solar system, OPEC, Big Oil, the nuclear industry and the utility companies would be howling with anguish.

The new OTEC-System, plus hot-water and radiant-heating solar, passive solar construction, and wind energy, would make the Western world totally independent of fossil fuels in our homes, and in most industrial plants and commercial buildings.

With plenty of OTEC-generated power, electricity would be so plentiful and inexpensive that the electric-powered car would most certainly come into its own.

National Plumbing Information Service, which I've mentioned frequently in this book, has compiled data generated from a number of reliable sources including the College of Engineering of the University of California, at Berkeley, in the person of Mr. Bryn Beorse, Research Engineer Emeritus of the Sea Water Conversion Laboratory. Mr. Beorse makes no bones about the fact that alternative energy is here to stay, and that if it is given a fraction of the funding which has been lavished upon nuclear research, new energy sources could be tapped to an enormous degree within our lifetimes—i.e.

the total energy requirements of America, plus a substantial reserve, could be covered, without recourse to petroleum or nuclear energy.

As with anything else good, the tendency has been to exaggerate merits and overlook faults. Certainly, that's been the history of solar hot-water heating.

Now, finally, a good, sound, workable system has emerged from the commercial chaos that had been fostered by one exploiter and con-person after another.

Solar was new, exciting, and ripe for promotion in the seventies, and of course the hucksters rushed into the breach that was opened up by the gas crisis, with scheme after scheme, designed to part the gullible from as much money as possible in the shortest amount of time.

Most of our present, effective solar systems are based upon Israeli experiments and designs. They were the first to exploit this most ancient source of boundless energy. Israeli science concentrated on it out of national economic necessity, with the incentive of having an abundant supply of the basic fuel in hand, for a substantial portion of each year.

Before I go into the construction of your new solar hot-water system, I'd like to mention the resources available to you for familiarization and, indeed, for practical assistance in connection with the sizing of your system, its design, and in the obtaining of advanced, well-designed products.

1. *Uniform Solar Energy Code* (1982 Edition, as amended, the International Association of Plumbing and Mechanical Officials (IAPMO), 5032 Alhambra Avenue, Los Angeles CA 90032 (approximately $10).

2. *The Alternative Energy Guide,* National Plumbing Information Service, Post Office Box 3455, San Rafael, CA 94912-3455 (approximately $9.50).

3. *The Next Whole Earth Catalog,* CoEvolution Quarterly, Post Office Box 428, Sausalito, CA 94966 (approximately $15).

4. *Solar Components Catalog,* Solar Components Corporation, Post Office Box 237, Manchester, NH 03105 (approximately $3).

5. *Popular Mechanics* and similar magazines offer alternative energy information in practically every issue. Recently, there have been series on new and innovative solar hot-water heating products

in so many national magazines that it would take far too much space to enumerate them all.

6. Yardbirds of California, 2360 Mendocino Avenue, Santa Rosa, CA 95401 (Attention: Mr. John Headley, President), are suppliers of affordable solar equipment of good quality.

Skeptic-friends of mine have often reminded me of the days of the huckstering of solar systems, and how abysmally many of them performed after people had paid many thousands of dollars to have them installed.

That is true, or rather it was true. Competition and the availability of highly developed, reliable, and much more efficient systems, manufactured and sold by reputable business enterprises, have brought about the emergence of new and effective solar products and techniques. However, *caveat emptor* (Let the buyer beware)!

PLANNING YOUR SOLAR HOT-WATER SYSTEM

Go outside and look at the roof of your house. If you have a flat or pitched surface to which you can attach two to four solar collectors of 4-ft. x 8-ft. or 4-ft. x 10-ft. sizes—that is, facing *south* or *approximately south*—you can have rather inexpensive, do-it-yourself solar hot-water heating.

A more-or-less standard three-bedroom, two-bath, contemporary American home, having one 40-gallon gas or 52-gallon electric hot-water heater, can probably be served by two 4-ft. x 8-ft. panels and an 80- to 90-gallon solar storage tank.

This presupposes an area in reasonably sunny latitudes.

Even at the higher North American latitudes, solar power can be effective, but it requires a bit more panel area, more and better insulated than normal storage, and a great deal more attention to panel orientation (the pitch and angle of the panels).

South of the border to the equator, solar could nearly eliminate the need for any other hot-water heating method, for most of the year. Sun-belt states have the same potential.

To give you an example: Don and Karen Laudenslager of Novato, California, have reduced their gas consumption to the point that

Orient Your Solar-Collector Panels to Face South

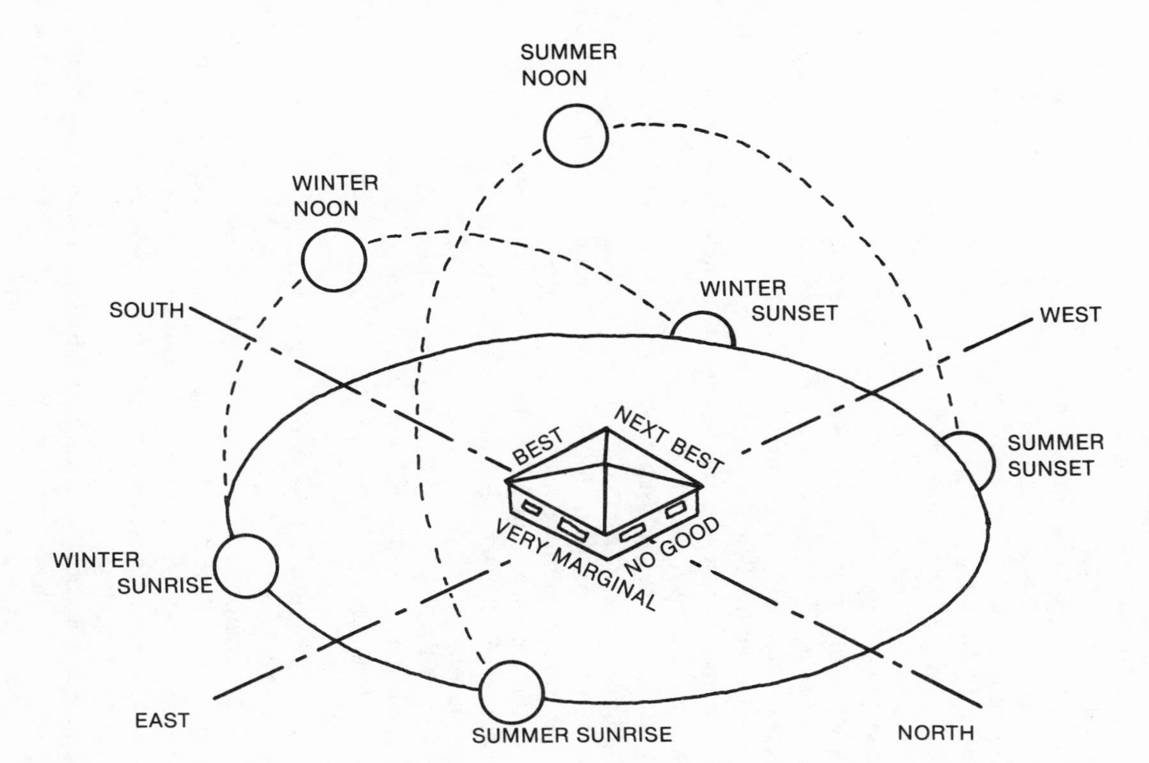

Square house, surrounded by a circle, through which centerlines have been drawn. Dotted lines join circles representing the transit of the sun over the house.

Don describes as "awe inspiring" from early spring to late fall. They have a somewhat larger tract-type contemporary home, served by two 4-ft. x 10-ft. collectors.

Now, I'm going to describe their system, as a model for yours.

Their two panels are facing generally *southwest.* Their original gas water heater is in a furnace room in the garage, which now holds an 82-gallon solar storage tank as well.

Water is circulated through the collectors by a 1/20-horsepower electric pump (which consumes the equivalent of a 60-watt light bulb, when it is operating).

They have a digital-readout, solid-state control, connected by sensors to the panels, to the storage tank, and to the solar-return piping where it enters the storage tank.

To regulate the flow through the solar collector, and drain down the collectors in case of a freeze, there is a dual-function automatic valve, operated by the electronic control.

Where the water is discharged from their gas hot-water heater into the house lines, there is a special tempering valve, designed to mix overheated solar hot water with cold water to provide a safe temperature for normal domestic use.

On the panels, I've installed three valves: (1) a temperature-and-pressure (T & P) relief valve, (2) a vacuum breaker, and (3) an automatic air vent.

Along with five gate valves, two hose bibbs, and a couple of mechanical temperature gauges, that is the basic system.

All of the piping is 3/4", type M and a limited amount of type L, on the roof.

Their system is insulated with pipe insulation designed for use in temperate climates.

You are going to have to insulate all of your lines with some sort of closed-cell material. Urethane of 7/8" wall thickness is suitable for inside the house. Outside takes 1" walls and a sun-resistant jacket. In warm latitudes, walls may be 1/2" for the inside, and 3/4" (with a jacket) for the outside.

You will also need to insulate your back-up gas or electric water heater, and I suggest an additional blanket of heater insulation around the solar tank as well. The more insulation, the less the heat loss.

BUYING MATERIALS

Benchmark names in solar energy are Grundfos, Sunspool, Grumman, Radco, Independent Energy, Heliotrope General, American Solarstream, Watts, Richdel, and Amtrol. There are many others, but I am not as familiar with them.

Go to your reputable solar-equipment store and ask qualified members of the staff for their recommendations. I have no doubt that you will be pointed in the right direction.

The only thing you need to know when you approach them with a view to buying a solar system, is that you are interested in the drain-down type of system, rather than the circulation-protected system.

The differences in initial and operating costs between the two types of systems are significant.

The drain-down type empties its collectors automatically, when there is a freeze.

The circulation-protected type circulates water through the collectors to keep them from freezing. You're pumping water when it can't possibly do you any good from the standpoint of making hot water. The initial cost of these types of systems is a bit more, usually.

Once you've seen the recommended design of the system you are investigating, you can list the pipe, valves, and fittings as if you were building a normal potable-water system, for which I've already given you ample information, in Sections Three and Five.

HELPFUL HINT: As you know, I am very concerned about safety on the job. That caution has resulted from quite a few broken bones, and a near brush with the grim reaper. Working on a roof, soldering pipe, is a damned dangerous business.

Recently, a product came out called the Sure Fire torch, which self-ignites via a trigger, thus making one-hand operation of the torch possible, and saving gas in the bargain.

For the economy-conscious among you, a standard torch is quite satisfactory. But, if you plan to do solar plumbing on your roof, or high off the ground, I strongly recommend the self-igniting Sure Fire torch.

If you can't find one of these torches at your local supply house, you can write to the manufacturer, Ignitor Products International,

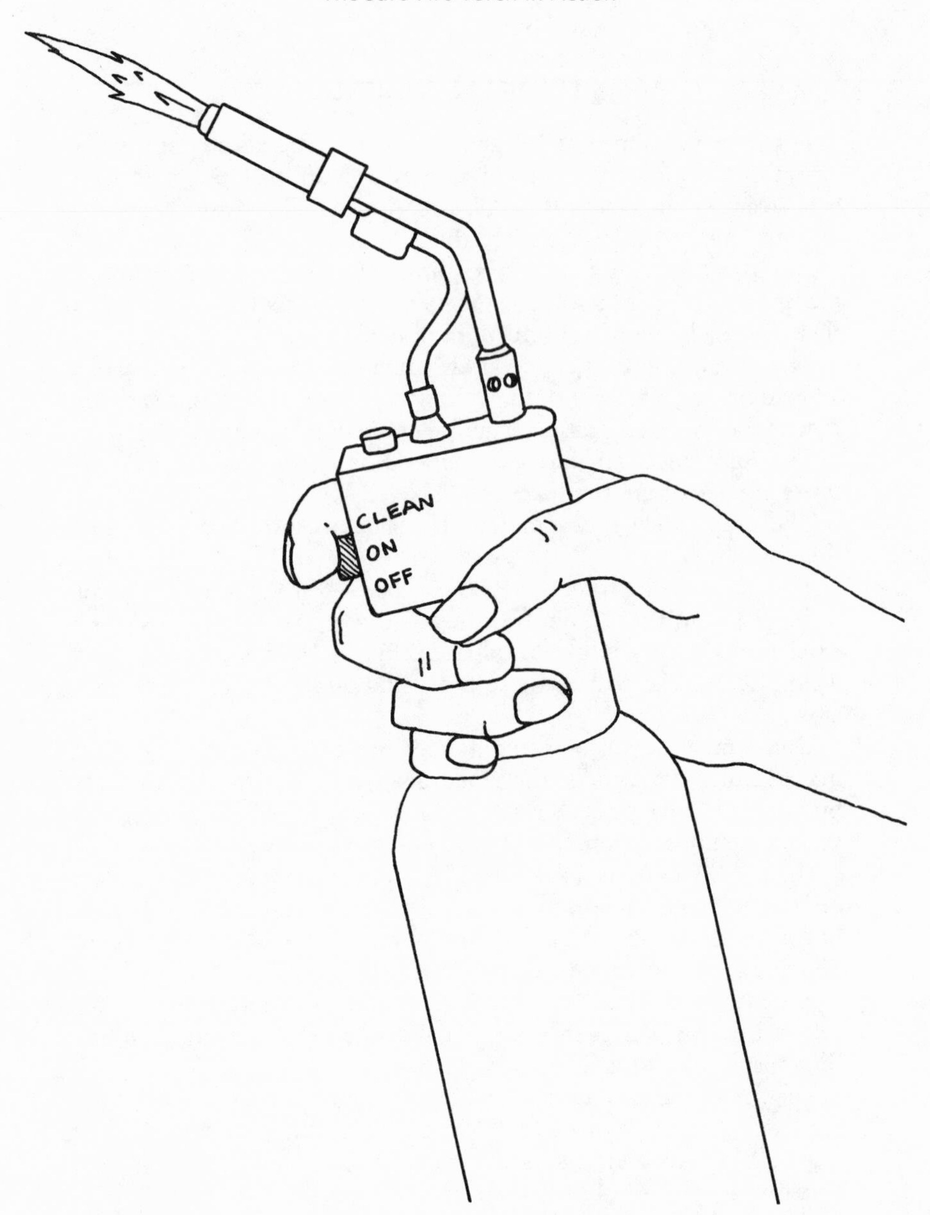

Hand holding Sure Fire torch

850 John Watts Drive, Nicholasville, KY 40356 (Attention: Mr. Benson Miller).

BUILDING YOUR SOLAR SYSTEM

Using the Laudenslagers' system as an example, the following items were installed: On the roof were two Radco 4-ft. x 10-ft. collectors, one Watts automatic air vent, one ¾" Watts vacuum-relief valve (vacuum breaker), one Watts water-heater temperature-and-pressure relief valve (with long-reach sensor rod), one temperature sensor, and two freeze sensors. The collectors were mounted on standard Olin brackets supplied by the manufacturer of the collectors. The upper brackets were mounted, in turn, on cadmium-plated steel struts, also furnished by the manufacturer, to elevate the panels to the recommended angle for this latitude (sun path) and the compass angle of the roof pitch.

Down in his furnace room, we installed an 82-gallon American Appliance Solarstream solar-storage tank equipped with a 240-volt electric standby element of 4,500 watts.

On to the tank we attached a Grundfos $\frac{1}{20}$-horsepower stainless-steel circulating pump, and a Sunspool drain-down valve, both of which connected to a Delta-T digital solid-state control. The Delta-T temperature sensor was substituted for the Solarstream sensor behind the lower access plate of the solar storage tank.

Copper lines were run to his old gas water heater (now called the backup heater), which delivers water to his house through an Amtrol tempering valve.

CAUTION: *Don't bypass this important component. The tempering valve can keep you and your family from being scalded by solar-heated water, which will be drawn into the backup heater instead of cold water. Solar-heated water can rise to over 190 degrees. The tempering valve mixes cold water with the solar-heated water to give you hot water at the temperature to which you've become accustomed.*

The system works by using the solar tank as the source of supply for your old water heater. Of course, if the solar-heated water is hotter than the setting on your water heater's thermostat, the gas will not turn on to start heating the water in the standby water heater. When the solar-heated water comes into your heater at a temperature less than your thermostat setting, the heater will fire up and

Solar-Equipment Room Layout

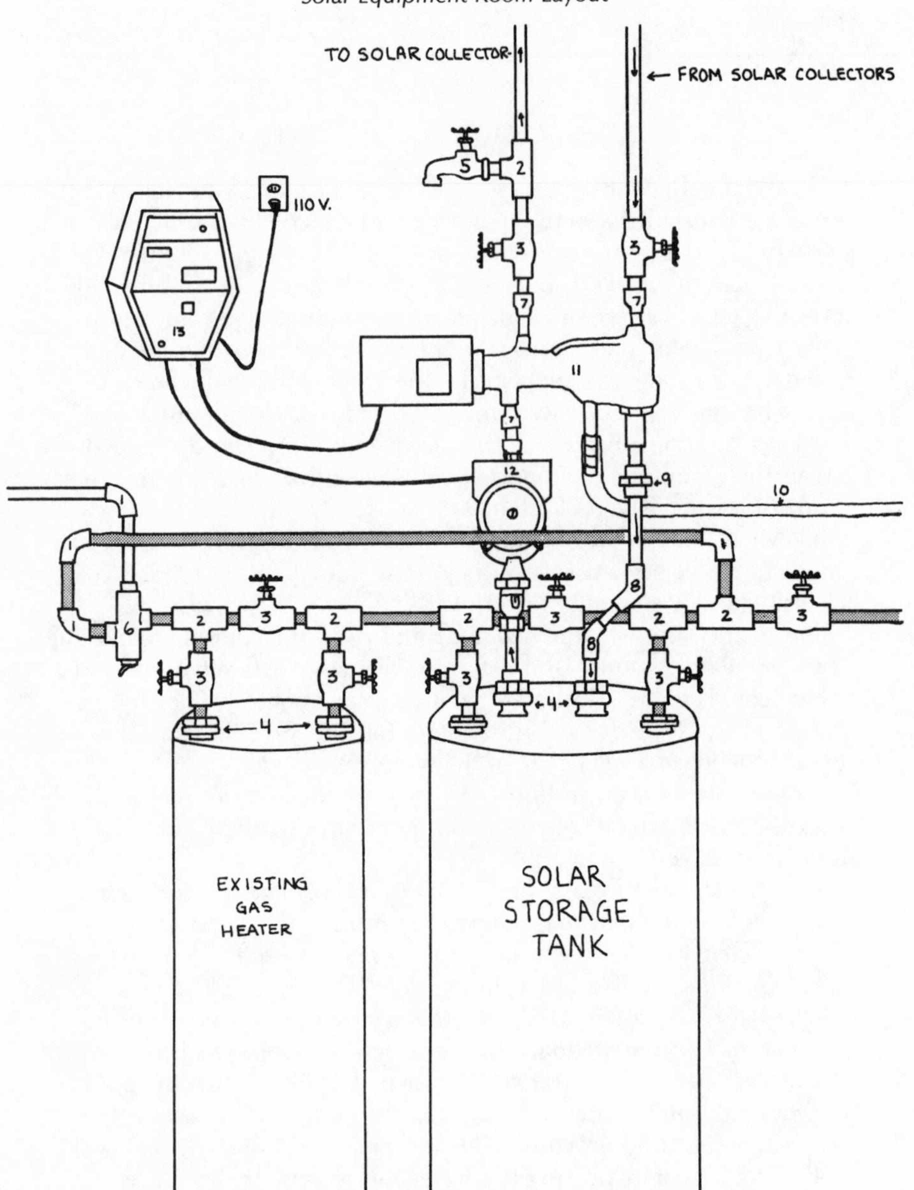

Solar-equipment room layout, including existing gas heater and solar storage tank

SOLAR FITTINGS LIST

1. 3/4" COPPER 90° ELBOWS
2. 3/4" COPPER T'S
3. 3/4" GATE VALVES
4. DIELECTRIC UNIONS
5. 3/4" HOSE BIBB
6. TEMPERING VALVE
7. 3/4" X 1/2" COPPER REDUCERS
8. 3/4" COPPER 45° ELBOWS
9. 3/4" UNION
10. 1/2" VINYL DRAIN-DOWN TUBE
11. SUNSPOOL DRAIN-DOWN VALVE
12. GRUNDFOS CIRCULATOR
13. DELTA T DIFFERENTIAL TEMPERATURE THERMOSTAT
14. 1" UNION
15. CAP (1")
16. 1" X 3/4" COPPER REDUCER
17. 1" X 3/4" COPPER T
18. AUTOMATIC AIR ELIMINATOR
19. VACUUM RELIEF VALVE
20. TEMPERATURE AND PRESSURE RELIEF VALVE
21. BLOW-OFF TUBE

FOR SENSOR PLACEMENT SEE "DELTA T" INSTRUCTIONS.

ALL SOLAR LINES MUST BE INSULATED TO LOCAL CODE

Solar equipment and fittings list.

Solar Panels, Piped and Fitted to Code

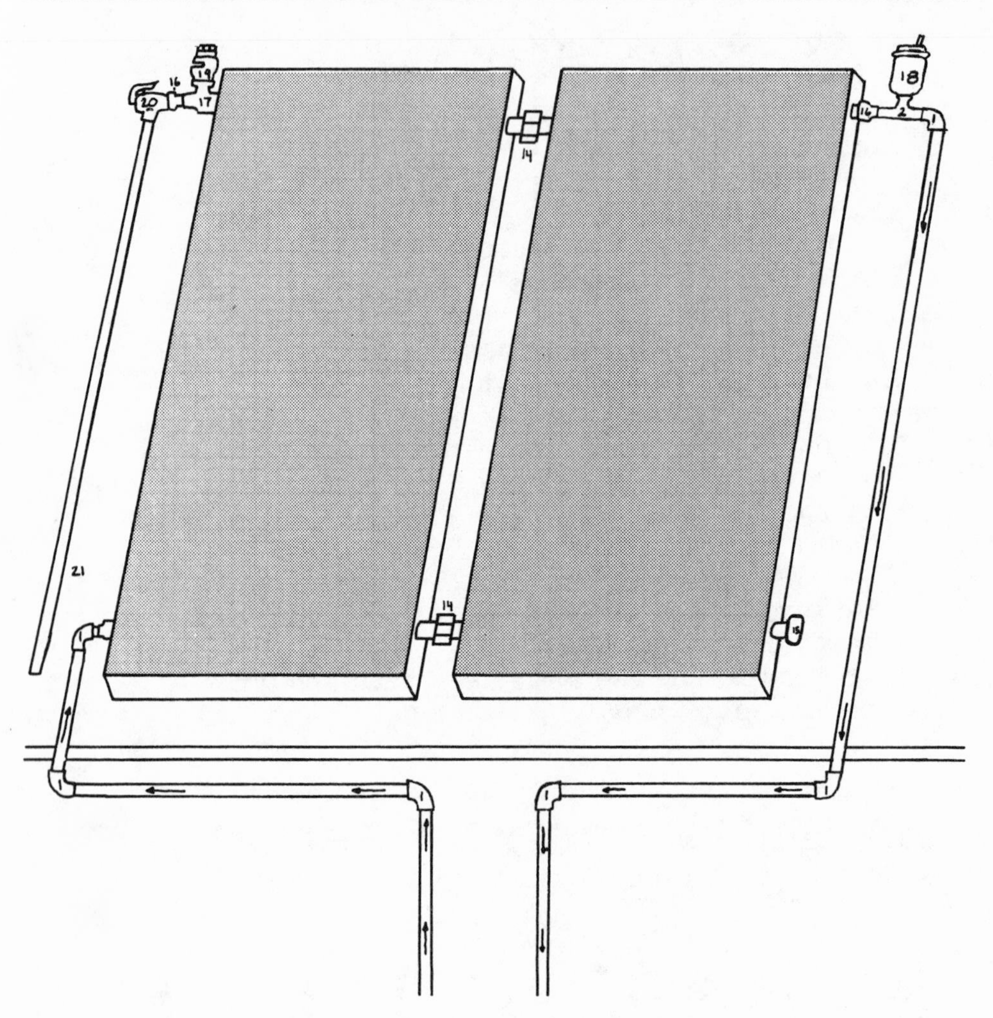

Two solar panels with piping and fittings in place.

Collector Mounting Brackets, Four Required Per Collector

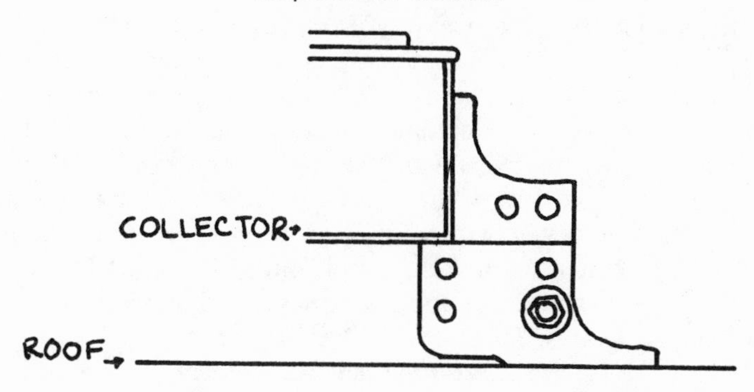

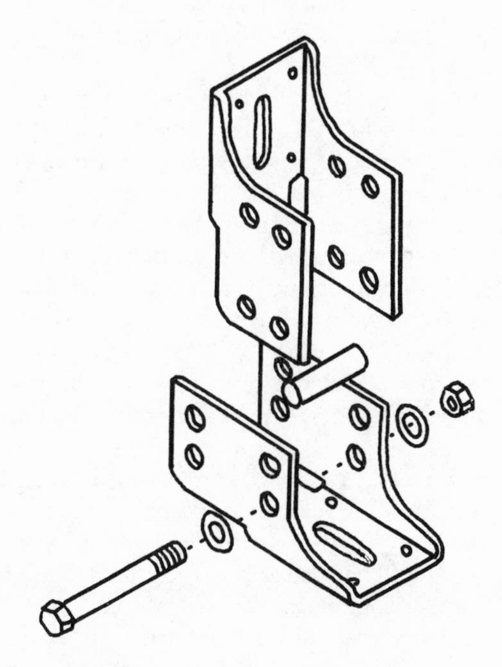

Standard mounting brackets for collectors.

heat the water the rest of the way. But even here there are savings. If you have any temperature increase in your heater's supply water, the gas will not have to stay fired up nearly as long to get it up to temperature. That will reflect in definite savings over the long haul.

The accompanying picture-diagram shows how your solar system should be installed, and where the valves and fittings go, between the solar tank and its components, and the panels on the roof. In this case, a picture is worth a thousand words.

Follow the diagram, use the principles of copper-pipe construction that I've described in Section Five, and get to it.

HELPFUL HINT: *The most important advice I can give you is to read the manufacturer's instructions on each piece of equipment included in your solar package, especially if you use the Sunspool and Delta-T combination.*

Be sure that you go through the installation of the equipment in detail with the salesperson at the store from which you purchase your components. Ask all the questions you can think of at the time, and don't hesitate to go back and ask more of them, if you should have any later.

Don't do anything, install any component, or run any pipe for any solar installation unless you are 100 percent sure how the pieces will fit together when you build, and how they'll work when the job is finished.

If you are not going to use the components named in the diagram, all you need to do is substitute those items which you've bought. The only thing that's important is function. Just make sure that each component you have purchased does exactly the same job as the parts I've shown.

Make sure your roof is sealed with some sort of bituminous or silicone sealing compound, where the collectors are bolted to the surface.

HELPFUL HINT: You'll save a whole bunch of heat if you insulate the cold-water line for three feet before it comes into the solar tank.

Using my diagram, plus the instructions which come with your components, you should have solar hot water for around $2,500.

In some states, like California, the utility companies are offering very low-interest loans, and the government is granting tax credits for solar installations that meet established norms of good design and construction.

And, you know what? You can do it yourself.

INDEX